Load Frequency Control of Microgrids

The book focuses on describing the emerging microgrid concept, and its various constituents, especially the EV technology, and investigates the load frequency control performance of different microgrid configurations by implementing the modern control theory. An exhaustive study is presented on the various renewable energy sources and an up-to-date status of their installed capacity and power generation. The text presents case studies for load frequency control of a microgrid in its various operating modes.

This book
- Discusses power system stability, significance of load frequency control in power system, modelling of a microgrid, and frequency deviation response.
- Covers various energy storage technologies such as superconducting magnetic energy storage, electrochemical energy storage, and compressed air energy storage.
- Presents modes of interaction of the electric vehicles with the utility grid and implementation of the electric vehicles for load frequency control.
- Illustrates control approaches for load frequency control and metaheuristic optimization algorithms.
- Showcases the study of load frequency control for standalone microgrid systems, grid-connected microgrid systems, and multi-microgrid systems.

It is primarily written for senior undergraduates, and graduate students in the fields of electrical engineering, electronics, communication engineering, and renewable energy.

Load Frequency
Control of Microgrids

Bhuvnesh Khokhar, K P Singh Parmar,
Tripta Thakur and D P Kothari

CRC Press
Taylor & Francis Group
Boca Raton London New York

CRC Press is an imprint of the
Taylor & Francis Group, an **informa** business

ISBN: 978-1-032-71831-6 (hbk)
ISBN: 978-1-032-76101-5 (pbk)
ISBN: 978-1-003-47713-6 (ebk)
ISBN: 978-1-032-76420-7 (eBook+)

DOI: 10.1201/9781003477136

Typeset in CMR10
by KnowledgeWorks Global Ltd.

This work is dedicated

To my Parents, wife Rimpy, and son Reyansh

Dr. Bhuvnesh Khokhar

To my Parents, wife Atri, daughter Vaidehi, and son Yagyesh

Dr. K P Singh Parmar

To all my Family Members

Dr. Tripta Thakur

To my wife Shobha

Dr. D P Kothari

Contents

Preface

Over the course of nearly two decades, microgrids (MGs) have emerged as a prospective alternative to conventional power systems reliant on fossil fuels. The cause of this phenomenon can be attributed to the rapid depletion of fossil fuel reserves and the adverse environmental impacts resulting from emissions generated by the conventional power systems. On the other hand, the MGs depend on renewable energy sources (RESs) that are readily accessible and environmentally friendly for the purpose of generating power. This elucidates the underlying factors contributing to their swift emergence. However, because to the variable characteristics of the RES-based generators (like solar photovoltaic array, wind turbine generator, etc.), the MGs face certain stability issues.

This book provides a concise description of the various categories of power system stability as classified by the IEEE PES task force, with an emphasis on the frequency stability problem arising in an MG. Since the fluctuating power output from RES-based generators has a direct impact on the generation-load balance, thereby altering the system frequency, it is evident that the stated problem warrants a greater level of investigation. As a result, stability, security, and reliability of the system are compromised. Secondary frequency control, or more commonly load frequency control (LFC), is the process of maintaining system frequency and tie-line power deviations within predetermined limits. Various MG configurations are mathematically modeled and evaluated for their LFC performance enhancement by implementing various prevalent modern control strategies. Thus, efforts have been made to reconcile the gap between effective MG frequency control and contemporary control theory.

Although the fossil fuels like coal, oil, and natural gas have been the driving force behind the energy boom in the recent centuries but at the same time rapid depletion of their deposits calls for alternative sources of energy, i.e., the RESs, that tend to be available for supply for a long long time and operate in an environmental friendly manner. Consequently, this book focuses on the current status and future trends of the diverse RESs in India as well as on the global platform in an exhaustive manner. In parallel, opportunities of the RESs for a sustainable development are also discussed along with the probable challenges in their integration.

Further, the aim of this book is centralized to acutely taking up the concept of an MG, its modes of operation, and classification and simultaneously examining its various subsystems i.e., distributed generation (DG) sources and energy storage (ES) systems along with their basic operating principles, classifications, advantages, and limitations. Considering the several inherent advantages related to the electric vehicles (EVs) be it economical, technical, environmental or customer convenience, a separate chapter is dedicated to the study of the EV technology that covers its classification, modes of interaction with the utility grid, implementation for the LFC, current status and future trends from both the Indian and global perspectives, and the dominant barriers identified in their deployment.

In addition, the purpose of this book is to examine the concept of an MG, its modes of operation, and its classification, as well as its various subsystems, including distributed generation (DG) sources and energy storage (ES) technologies, as well as their fundamental operating principles, classifications, advantages, and limitations. Considering the several inherent advantages related to the electric vehicles (EVs) be it economical, technical, environmental or customer convenience, a separate chapter is dedicated to the study of the EV technology that covers its classification, modes of interaction with the utility grid, implementation for the LFC, current status and future trends from both the Indian and global perspectives, and the dominant barriers identified in their deployment.

The book is divided into nine chapters. **Chapter 1** introduces the diverse classifications of power system stability, emphasizes the importance of the LFC in power systems, and explores the different frequency control approaches employed in them. The present discussion delves into the importance of frequency stability as perceived from the perspective of an MG. Additionally, a comprehensive outline of the general framework of an MG is provided along with discussion on the dynamics of a case study MG.

Chapter 2 explores the significance of the RESs in the context of achieving sustainable development. It investigates the diverse opportunities associated with the RESs and highlights the advantages they offer. This chapter further presents an analysis of the existing state and future projections of the RESs for power generation, focusing on both Indian and global contexts. Moreover, the chapter includes a comprehensive classification of the RESs and an examination of the potential obstacles associated with their adoption.

Chapter 3 provides a comprehensive analysis of the several DG sources and ES technologies that are now available. This chapter examines the operational concepts, classifications, benefits, and constraints associated with these technologies.

Chapter 4 provides an in-depth description of the future of the automobile industry, specifically focusing on the emerging EV technology. The chapter presents a classification of the EVs and discusses the many modes of integration between EVs and the utility grid. Further, this chapter explores the importance of the EVs from the perspective of the LFC. It also analyses the current state and future developments of EV deployment on a global scale, as well as within the context of India. Moreover, the significant policies aimed at promoting the deployment of the EVs and overcoming the prevailing barriers in this domain are presented.

Chapter 5 provides a comprehensive examination of the diverse standard definitions pertaining to an MG, along with its composition, modes of operation, and classification. This chapter provides a comprehensive analysis and discussion of the many DG sources and ES technologies.

Chapter 6 highlights the importance of several control approaches for LFC, as well as their classification, benefits, and limitations. The chapter also presents an overview of the theory pertaining to metaheuristic optimization algorithms (MOAs), including its general structure, classification, applications in addressing various power system issues, as well as their advantages and limitations. Additionally, this chapter presents the mathematical derivation of the performance indices that are often utilized.

Chapters 7, 8, and 9 the LFC performance of a standalone MG, a grid-connected microgrid (GcMG), and a multi-microgrid (MMG) system are analyzed separately, and the mathematical models for each configuration are described in detail. We investigate the controller models that are implemented as the secondary controllers in each of the setups, as well as the steps that are involved in optimizing the parameters of those controller models using the various MOAs. In the final step, the time-domain simulation results are carefully analyzed, taking into consideration each of the modeled MG configurations.

Academic scholars and engineers conducting research on MG operation and control may find this book useful for their research. It can be beneficial as an additional input for electrical engineering students attending university courses on renewable energy, EVs, power system stability, operation, and control at the graduate, postgraduate, and research levels, as well as for power system operators.

About the Authors

Dr. Bhuvnesh Khokhar is currently Assistant Professor, Department of Electrical Engineering with Galgotias College of Engineering and Technology (GCET), Greater Noida, Gautam Buddha Nagar, Uttar Pradesh, India. He obtained his BE (Hons) in Electrical Engineering from Chhotu Ram State College of Engineering (now Deenbandhu Chhotu Ram University of Science and Technology), Murthal, Sonipat, Haryana, India in 2010, M.Tech (Hons) in Power Systems in 2012 and Ph.D in Power Systems in 2021 from Deenbandhu Chhotu Ram University of Science and Technology, Murthal, Sonipat, Haryana, India. His research areas include operation and control of microgrids, renewable energy, electric vehicles, economic load dispatch, optimization algorithms, intelligent, and robust control. Dr. Khokhar has published around 20 research papers in various National and International Journals and Conference Proceedings including *Applied Energy* (Elsevier), *Applied Soft Computing* (Elsevier), *Arabian Journal for Science and Engineering* (Springer), and *Electric Power Components and Systems* (Taylor & Francis).

Dr. K. P. Singh Parmar is a Deputy Director (Technical) with NPTI, Faridabad, Haryana, India. He obtained his BE (Hons) in Electrical Engineering from GEC, Rewa (MP), M.Tech. in Energy from IIT, Delhi and Ph.D. in Electrical (Power Systems) from IIT, Guwahati. He has been involved in teaching, training, consultancy, and research since 2001. His contributions to the consultancy assignment of setting up a National Power Academy in the Kingdom of Saudi Arabia (KSA) have been well appreciated. He was the project in-charge of DPR preparation for conversion of HT/LT overhead lines to Underground cables, automation, etc., for the Ayodhya city under MVVNL. He has contributed about 50 research papers in the National/International Journals and Conference Proceedings. He has contributed a chapter on AC Machines included in a book published by IGNOU, New Delhi. He has also authored a book on Management of Transmission System. Recently, Dr. Parmar has contributed a book chapter in a book titled Advanced Frequency Regulation Strategies in Renewable Dominated Modern Power Systems (Elsevier) 2024.

His research interests include Automatic Generation Control, Optimal load dispatch/scheduling, Power System Operation and Control, Power System Restructuring, Renewable Energy, Energy Management System, AI Applications in Power System.

Dr. Tripta Thakur is Director General, National Power Training Institute (NPTI), apex body of Ministry of Power, Government of India. She was earlier Head and Professor, Electrical Engineering Department at the National Institute of Technology, MANIT-Bhopal, India. She is a graduate in Electrical Engineering with Master's degree in Power Electronics from IIT-Kanpur, and has a PhD from IIT-Delhi. She has been recipient of several awards such as Commonwealth Research Scholar at University of Dundee (2005-2008), UK, Commonwealth Academic fellow at Durham University Business School (2014), UK, COFUND Senior researcher at Durham University Business School (2016), Visiting Faculty at Asian Institute of Technology, Bangkok (2010), technical member for International Electrotechnical Commission (IEC), SEG4 Group, ISGF (MoP) working group member, etc. Recently, she received the ISGF Innovation Awards 2024 for the Category "Women in Power". She has teaching and research experience of 28 years and has nearly 100 publications to her credit. She has also been a Consultant for evolving a possible Common South Asian Electricity Markets. She has done various consultancies for Distribution companies in India.

Dr. D. P. Kothari obtained his BE (Electrical) in 1967, ME (Power Systems) in 1969 and Ph.D in 1975 from BITS, Pilani, Rajasthan, India. From 1969 to 1977, he was involved in teaching and development of several courses at BITS Pilani. Earlier Dr. Kothari served as Vice Chancellor, VIT, Vellore, Director in-charge and Deputy Director (Administration) as well as Head in the Center of Energy Studies at IIT, Delhi and as Principal, VRCE, Nagpur. He was visiting professor at the Royal Melbourne Institute of Technology, Melbourne, Australia, during 1982-83 and 1989, for two years. He was also NSF Fellow at Perdue University, USA in 1992.

He has published/presented 840 research papers in various National as well as International journals, conferences, guided 57 Ph.D scholars and 68

M. Tech. students, and authored 75 books. Having received 89 awards till now, his awards include the National Khosla Award for Lifetime Achievements in Engineering (2005) from IIT Roorkee, UGC National Swami Pranavandana Saraswati Award (2005), etc. He is a Fellow of the National Academy of Engineering, Fellow of Indian National Academy of Science, Fellow of Institution of Engineers, Fellow IEEE, Hon. Fellow ISTE and Fellow IETE.

Currently, Dr. Kothari is Chairman, Board of Governors of THDC IHET, Tehri, India.

Acknowledgements

The majority of the information, findings, discourse, and perspectives expounded in this book were attained from the extensive expertise of the authors in teaching, training, research, and consultancy within the power sector domain. The authors express their gratitude for the support received from multiple sources, such as the National Power Training Institute (NPTI), Galgotias College of Engineering and Technology (GCET), Greater Noida, Powergrid, Central Transmission Utility (CTU), Grid-India, National Thermal Power Corporation (NTPC), National Hydro Power Corporation (NHPC), and Central Electricity Authority (CEA).

The authors would like to express their sincere gratitude for the invaluable support and contributions provided by numerous persons throughout the course of this endeavor. They express their obligation to Mr. Gauravjeet Singh Reen and the rest of the team at CRC Press/Taylor & Francis Group, together with Ms. Mehnaz Hussain for their contributions in editing, layout, and facilitating the publication of this book. The authors would like to extend their sincere appreciation to those individuals who have generously offered their support, supplied valuable comments, granted permission to them to use their views, and contributed to the editing, proofreading, and design aspects of this publication. Quantifying the exact number of individuals who have provided assistance in the authors' endeavor proves to be a challenging task; yet, a substantial portion of these individuals can be identified within the reference section of this book.

The authors would like to express their sincere appreciation to the officials of the Ministry of Power, Government of India, for their helpful cooperation and guidance in completing this book. They also express their appreciation to the officers, faculty members, and trainees at the NPTI for their outstanding contributions in improving the quality of this book. The authors express their gratitude to the management and professors of the GCET, Greater Noida for their important contributions and recommendations.

Ultimately, the authors express their sincere appreciation to their respective families for their unwavering support and invaluable assistance during the process of compiling this book.

Bhuvnesh Khokhar
K P Singh Parmar
Tripta Thakur
D P Kothari

List of Figures

List of Tables

Abbreviations

AGC	automatic generation control
BESU	battery energy storage unit
BEV	battery electric vehicle
CAES	compressed air energy storage
CB	circuit breaker
CEA	Central Electricity Authority
CHP	combined heat and power
CIG	converter interfaced generation
CSA	crow search algorithm
CSP	concentrated solar power
DD	double derivative
DG	distributed generation
DISCOM	distribution company
DoF	degree-of-freedom
ESO	energy storage obligations
ESU	energy storage unit
EV	electric vehicle
FACTS	flexible AC transmission system
FC	fuel cell
FDR	frequency deviation response
FO	fractional order
GcMG	grid-connected microgrid
GDB	governor dead band
GM	gain margin
GRC	generation rate constraint
HVDC	high voltage direct current
IAE	integral of squared error
IC	internal combustion
IEA	international energy agency
IMC	internal model control
IRENA	international renewable energy agency
ISE	integral of squared error
ITAE	integral of time multiplied absolute error
ITSE	integral of time multiplied squared error
LDC	load despatch center
LFC	load frequency control
LPF	low-pass filter
MG	microgrid
MMG	multi-microgrid

MOA	metaheuristic optimization algorithm
MPC	model predictive control
MT	micro-turbine
PCC	point of common coupling
PEC	power electronic converter
PHEV	plug-in hybrid electric vehicle
PID	proportional integral derivative
PLP	pulse load perturbation
PM	phase margin
PV	photovoltaic
RES	renewable energy source
RFBU	redox flow battery unit
RLP	random load perturbation
RTC	round-the-clock
RTPP	reheat thermal power plant
SCA	sine cosine algorithm
SLP	step load perturbation
SMES	superconducting magnetic energy storage
SoC	state-of-charge
SRAS	secondary reserve ancillary service
SSA	salp swarm algorithm
TD	time delay
TEM	total energy model
TRAS	tertiary reserve ancillary service
TWh	terra watt hour
UCU	ultra capacitor unit
V1G	vehicle-1-grid
V2C	vehicle-to-customer
V2G	vehicle-to-grid
V2H	vehicle-to-home
WTG	wind turbine generator

Nomenclature

$_AQ_t^{\tilde{\alpha}}$	fractional operator
a	proportional setpoint weight of the 2DoF-PID controller
A_r	area swept by the WTG blades (m^2)
b	proportional setpoint weight of the 2DoF-PID controller
B	bias factor of the MMG system
c	output signal of the secondary controller
C_p	power coefficient of the WTG
C_{kW}^*	inverter capacity of the EV energy storage system
C_{kWh}^*	battery capacity of the EV energy storage system
D	load damping coefficient of the MG (pu MW/Hz)
Δf	system frequency deviation
Δf_e	frequency deviation range for emergency control scheme
Δf_p	frequency deviation range for primary control scheme
Δf_s	frequency deviation range for secondary control scheme
$\Delta P_d / \Delta P_D$	change in load demand
ΔP_{DG}	change in the output power of the DG sources
ΔP_{DEG}	change in the output power of the DEG unit
ΔP_{ESU}	change in the output power of the ESU
ΔP_{EV}	change in the output power of the EV model
ΔP_g	change in the output power of generator
ΔP_m	change in the output power of turbine
ΔP_{MG}	change in power output of the MG
ΔP_{PV}	change in the output power of solar photovoltaic
ΔP_{RTPP}	change in power output of the RTPP
ΔP_s	change in the output signal of speed changer motor
ΔP_{SG}	change in power output of the SG unit
$\Delta P_{tie-line}$	change in tie-line power of the MMG system
ΔP_{WTG}	change in the output power of wind turbine generator
η	conversion efficiency of the PV unit
$E_{control}^{max}$	maximum capacity limit of the EV
$E_{control}^{min}$	minimum capacity limit of the EV
f_0	nominal system frequency
K_{AE}	gain of the AE unit
K_d	derivative gain of the secondary controller
K_{dd}	double derivative gain of the secondary controller
K_{DEG}	gain of the DEG unit
K_{FC}	gain of the FC unit
K_i	integral gain of the secondary controller
K_p	proportional gain of the secondary controller

K_{PS}	gain of the GcMG
K_R	gain of the reheater
λ	order of the integrator of the FOPID + DD controller
μ	order of the differentiator of the FOPID + DD controller
M	inertia constant of the MG (s)
N	derivative filter coefficient of the secondary controller
$N_{control}$	total number of controllable EVs
$N_{control-in}$	number of EVs transferring from charging state to controllable state
$N_{initial}$	initial number of controllable EVs
$N_{plug-out}$	number of EVs transferring from controllable state to driving state
ω_{gc}	gain crossover frequency (rad/s)
ω_{pc}	phase crossover frequency (rad/s)
OF_{ISE}	ISE based objective function
OF_{ITSE}	ITSE based objective function
ϕ_{solar}	solar radiation measured ($Watt/m^2$)
ρ	air density (kg/m^2)
R	governor speed regulation coefficient (Hz/pu MW)
S	surface area of the PV unit (m^2)
SoC_{avg}	average SoC of the energy storage system of the controllable EVs
t_d	derivative time of the conventional PID controller
t_i	integral time of the conventional PID controller
t_{max}	maximum number of iterations
T_a	ambient temperature (^oC)
T_{AE}	time constant of the AE unit (s)
T_{BESU}	time constant of the BESU
T_{DEG}	time constant of the DEG unit (s)
T_{EV}	time constant of the aggregate EV model (s)
T_{FC}	time constant of the FC unit (s)
T_{FESU}	time constant of the FESU
T_G	time constant of the governor (s)
T_{IC}	time constant of the inter connector
T_{IN}	time constant of the inverter
T_{PV}	time constant of the solar PV unit (s)
T_{PS}	time constant of the GcMG (s)
T_R	time constant of the reheater (s)
T_T	time constant of the turbine (s)
T_{WTG}	time constant of the WTG unit (s)
V_w	wind velocity (m/s)

1 Load Frequency Control: Significance, Definition, and Underlying Concept

The purpose of this chapter is to demonstrate the importance of load frequency control (LFC) in power systems. Following that, definitions and fundamentals of several types of power system stability are presented. Various frequency control schemes implemented in the power system to ensure its stability and reliability are reviewed. Further, the microgrid (MG) perspective of the LFC is italicized. Finally, frequency deviation responses (FDR) of a case study MG are comprehended along with a discussion on a generalized MG model.

1.1 POWER SYSTEM STABILITY

The operating conditions in power systems change continually whether they be related to a gradual change in load demand or generator power output. Concurrently, the power system is subjected to a wide range of disturbances from small to large. Small disturbances may be switching on/off of loads or tripping of a small generator, whereas large disturbance may be a short circuit on transmission line or loss of a large generator. In such a regime, stability of the power system may be impacted that needs to be tackled failing which the power system may transit into an unstable state. The IEEE/CIGRE joint task force in its report in 2004 redefined the definition of power system stability stating that the earlier definitions did not completely consider industry needs, experiences, and understanding [1]. The proposed definition was

> *Power system stability is the ability of an electric power system, for a given initial operating condition, to regain a state of operating equilibrium after being subjected to a physical disturbance, with most system variables bounded so that practically the entire system remains intact.*

In the report, the power system stability was classified into three types: (i) rotor angle stability, (ii) voltage stability, and (iii) frequency stability. In 2016, another IEEE PES joint task force was devised that reclassified the power system stability stating the fact that due to the increased penetration of converter interfaced generation (CIG) technologies, loads, and transmission devices, the dynamic behavior of the power system has gradually changed over the past years [2]. The CIG technologies include wind and solar photovoltaic

DOI: 10.1201/9781003477136-1

1

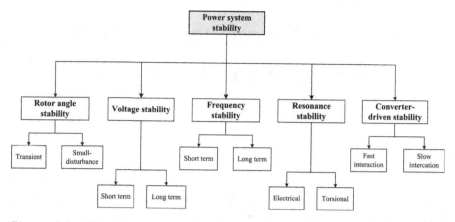

Figure 1.1 Classification of the power system stability as per IEEE PES 2020 report

(PV) power generation, energy storage technologies, flexible ac transmission systems (FACTS), high voltage direct current (HVDC) systems, lines, and power electronic interfaced loads. As per the report, the power system stability was reclassified as shown in Fig. 1.1. Definitions and fundamentals of each type of the stability are presented in brief below. For a detailed description, interested readers may refer to [2].

1.1.1 ROTOR ANGLE STABILITY

Rotor angle stability refers to the capability of the various interconnected synchronous machines in the power system to operate in synchronism under normal operating conditions, and if subjected to any small or large disturbance, they must be able to regain synchronism. To remain in synchronism, a machine must produce an electromagnetic torque equal and opposite to the mechanical torque delivered by its prime mover. As far as the CIGs replacing the conventional synchronous machines are concerned, it will result in an overall reduction in the total inertia of the power system. This ultimately would impact the rotor angle stability of the power system consequently changing the tie-line power flows, affecting the damping torque of the synchronous machines in vicinity, and replacing the synchronous generators equipped with vital power system stabilizers.

The presence of insufficient or negative synchronizing torque can give rise to the non-oscillatory instability known as transient instability. In synchronous machines, this particular sort of instability is what causes considerable rotor angle fluctuations, and it is frequently researched using approaches that involve numerical integration. On the other hand, the absence of a negative damping torque results in oscillatory small-disturbance stability. This is the case when the disturbances are of a sufficiently small magnitude. Following a system disturbance or a change in the system topology, this type of instability is symbolized by a complex conjugate pair of comparatively poorly damped eigenvalues of the linearized system state matrix shifting through the left-half plane (stable region) to the right-half plane (unstable region) of the complex plane.

1.1.2 VOLTAGE STABILITY

Voltage stability refers to the capability of a power system to hold the voltages steady at all the buses in the system as close as possible to the specified value after being subjected to a disturbance. Voltage instability in the power system may result in loss of load or loss of synchronism of generators in a control area and tripping of transmission lines or other network components by their respective protection schemes. Definition of voltage stability as defined in [2] is alike as defined in [1] except that the latter did not consider the impact of HVDC links on it. Terminals of the HVDC links with line commutated converters may result in the voltage instability. However, the use of voltage source converters in the HVDC converter stations has significantly improved the voltage stability limits.

In order to maintain short-term voltage stability, fast-acting load components including induction motors, electronically controlled loads, HVDC links, and inverter-based generators must be dynamic. The timescale of the study that is of particular interest is on the order of several seconds, much like that of rotor angle stability or converter-driven stability.

Equipment with a delayed response time is required to maintain long-term voltage stability. Examples of such equipment are tap-changing transformers, thermostatically controlled loads, and generator current limiters. In most cases, it takes the shape of a gradual decrease in the voltages that are present at various network buses. When some of the generators reach their field and/or armature current time-overload capability limits, this further restricts the maximum power transfer and voltage support that can be provided by those generators. The period of interest in the study could last for many minutes, and in order to evaluate the dynamic performance of the system, you need to run simulations over a prolonged length of time.

1.1.3 FREQUENCY STABILITY

Frequency stability is the capability of a power system to maintain steady system frequency after being subjected to a disturbance, resulting in a considerable mismatch between power generation and load demand. In general, frequency stability issues are an outcome of inadequacies in equipment responses, poor coordination of control and protection equipment, or insufficient generation reserves. Over the past two decades, various severe power grid blackouts have occurred globally resulting from the frequency instability [3]. These are listed in Table 1.1.

Table 1.1

List of power grid blackouts globally in the past two decades (in ascending order of date)

Country name	Year of blackout
Brazil	March to June 1999
Iran	Spring 2001 and Spring 2002
Northeast USA – Canada	August 2003
Southern Sweden and Eastern Denmark	September 2003
Italy	September 2003
Russia	May 2005
Europe	September 2006
Brazil and Paraguay	November 2009
India	July 2012
Thailand	May 2013
Turkey	March 2015
Java	August 2019
Argentina, Paraguay, and Uruguay	June 2019
Venezuela	March 2019
Sri Lanka	August 2020
Pakistan	January 2021

When considering the impact of the CIGs on the system frequency, studies show that the CIGs can contribute decisively and effectively in controlling the frequency, especially if the energy storage technology is simultaneously considered. It is noteworthy here that the low inertia of the CIGs, as compared to the conventional generators, results in continuous frequency excursions that may develop premature frequency instability problems in the power system. Consequently, this demands designing appropriate fast-acting controllers that can prohibit frequency deviations in the system as soon as they are detected. However, recent research has demonstrated that the frequency response of systems including the CIGs is a complicated phenomenon that calls for additional examination.

1.1.4 RESONANCE STABILITY

Resonance stability refers to the capability of a power system to restrain voltage/current/torque within specified limit after being subjected to large oscillations related to insufficient dissipation of energy in the flow path. The resonance, in general, is an outcome of a periodic energy exchange in an oscillatory manner. It is to be noted here that the basic concept of the resonance stability incorporates subsynchronous resonance (SSR) which is defined as [4]

A condition in the power system where the electric network exchanges energy with a turbine generator at one or more of the natural frequencies of the combined system below the synchronous frequency of the system.

According to the original publications on this phenomenon, the term SSR can appear in one of two ways: (i) as a result of a resonance between series compensation and the mechanical torsional frequencies of the turbine-generator shaft, or (ii) as a result of a resonance between series compensation and the electrical characteristics of the generator. The first of these occurs between the mechanical modes of torsional oscillations on the turbine-generator shaft (torsional resonance) and the series compensated electrical network, whereas the second is an entirely electrical resonance and termed as induction generator effect (IGE).

Reference [1] did not consider this stability aspect in its report but the fact that emergence of the FACTS technology and HVDC system and existence of the doubly-fed induction generators (DFIG) in the power system may impact the subsynchronous oscillations is the principal reason that this stability aspect has been newly reported in [2].

1.1.5 CONVERTER-DRIVEN STABILITY

Converter-driven stability is the capability of a power system to overcome unstable power system oscillations covering a wide frequency range (from less than 10 Hz to kHz), resulting from dynamic interactions of the control systems of the power electronics-based systems with components of the power system. The converter-driven instability problem may arise either due to fast dynamic interactions between the CIGs, HVDC systems or FACTS, and the transmission network or the stator dynamics of the synchronous generators or due to slow dynamic interactions between the former and electromechanical dynamics of synchronous generators or some generator controllers.

Fast-interaction converter-driven stability involves across-the-system stability issues resulting from rapid changing interactions between the control systems of power electronic-based systems (such as CIGs, HVDC, and FACTS) and the fast-response components of the power system, such as the transmission network, the stator dynamics of synchronous generators, or other power electronic-based devices. On the other hand, the slow-interaction converter-driven stability pertains to across-the-system instabilities caused by the gradual interactions between control systems of power electronic-based devices and the sluggish components of the power system, such as the electromechanical dynamics of synchronous generators and certain generator controllers.

1.2   SIGNIFICANCE OF LFC IN POWER SYSTEM

In light of the fact that the primary objective of this book is to show and investigate the LFC performance of an MG, prior to delving into the specifics, it is essential to comprehend the significance of the LFC in the power system. After going over some of the more fundamental aspects of frequency stability in the section that came before this one, the purpose of this section is to go over, in further detail, why it is essential for the power system to have frequency control. In a power system, the degree to which there is a mismatch between the amount of power generated and the amount of load demand has a significant impact on the system's reliability and stability. A persistent gap between these two can cause unwelcome and significant fluctuations in the system frequency, which can, in turn, damage expensive equipment, reduce load performance, overload transmission lines, and even result in a total blackout of the power system [3]. In such a scenario, the LFC plays an essential part by keeping the power generation in sync with the load demand and, as a result, preserving a steady frequency profile. In addition, the scale of the power system is growing, its structure is shifting, and various distributed generating technologies are being integrated, which calls for the implementation of an efficient LFC approach. Additionally, the operation of these technologies is unpredictable and intermittent.

1.3   FREQUENCY CONTROL SCHEMES IN POWER SYSTEM

Depending on the range within which the frequency in power system may deviate, three frequency control schemes can be formulated [3, 5]. These are: (i) primary frequency control, (ii) secondary frequency control, and (iii) emergency or tertiary frequency control. Figure 1.2 shows these control schemes with f_0 as the nominal system frequency, Δf_p, Δf_s, and Δf_e as the frequency deviation ranges for the primary, secondary, and emergency control schemes, respectively.

1.3.1   PRIMARY FREQUENCY CONTROL

The primary frequency control scheme is implemented when small frequency deviations occur in the power system. Here, the primary frequency control loop of the generator activates and responds within a few seconds. The system frequency is restored to nominal value through the droop action of the generator. Droop of a generator is defined as the ratio of change in frequency to the change in power output of the generator.

The primary control loop shown in Fig. 1.3 performs the primary frequency control where a change in the real power output (ΔP_g) of the generator, following a change in load (ΔP_d), is controlled by the mechanical power output (ΔP_m) of a turbine coupled to the generator. In case of a steam or hydro turbine, the ΔP_m is controlled by the steam (or water) input to the turbine

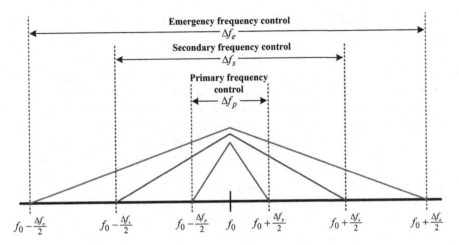

Figure 1.2 Various frequency control schemes in power system

through a valve gate. Following a load change, the turbine speed also varies and hence the system frequency. This speed variation is sensed by a speed governor present in the primary control loop which in turn directs a hydraulic amplifier to regulate the steam (or water) input to the turbine. Finally, a speed changer motor specifies the power output setting of the turbine. It is noteworthy here that the primary frequency control scheme is solely not efficient to restore the system frequency, especially for an interconnected power system. In that case, an additional, secondary load frequency control scheme is implemented to control the speed changer motor in accordance with load change in the power system.

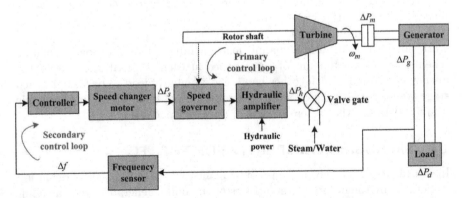

Figure 1.3 Primary and secondary frequency control loops

1.3.2 SECONDARY FREQUENCY CONTROL

For larger frequency deviations in the power system, the secondary frequency control is implemented. The secondary frequency control more commonly referred to as the LFC comes into action after the abnormal frequency deviations persist in the power system for more than a few minutes. The LFC automatically regulates the power output of the generator in accordance with the load variation and thus intends to confine the frequency deviations within specified limits and simultaneously, maintain the power exchange within different control areas at scheduled values. A more broader term generally used interchangeably with the LFC (including this book also) is the automatic generation control (AGC). Apart from performing all the functions of the LFC, the AGC additionally includes distributing the desired power output change among the different generators in order to minimize their total operating cost (termed as economic load dispatch).

The secondary frequency control loop is shown in Fig. 1.3. The loop implements a frequency sensor that senses the deviation in the system frequency and feeds it to a dynamic feedback controller. The resultant signal (ΔP_s) after passing through the speed governor and hydraulic amplifier, indirectly controls the output power of the turbine (ΔP_m) to track the load change and thus, restores the system frequency back to the nominal.

1.3.3 EMERGENCY FREQUENCY CONTROL

In case of a serious and persistent frequency deviation problem, following a severe fault condition in the power system, the LFC scheme may be unable to restore the system frequency back to the nominal value. In such a case, the emergency frequency control scheme is activated, such as load shedding, to prevent any cascade faults in the power system. The load shedding is implemented only when the system frequency drops below a threshold value.

The load shedding scheme, in general, can be classified into two types [3]: static load shedding and dynamic load shedding. In static load shedding scheme, a constant amount of load is shed at each stage till the system frequency is restored. On the other hand, the amount of load to be shed, at each stage, varies in accordance with the magnitude of disturbance and dynamic characteristics of the system in case of dynamic load shedding.

1.4 MICROGRID PERSPECTIVE OF THE LFC

In contrast to large conventional power plants, where the frequency deviation is dictated by the droop characteristics of the bulky generators with a significant quantity of inertia, there is no such provision available with the MGs. Further, due to the absence of direct coupling between the numerous DGs and the MG, the latter possesses a relatively low moment of inertia. An MG also makes it possible to enhance the penetration of RES-dependent units, such as solar PV or WTG units. These kinds of equipment operate intermittently due to their susceptibility to weather conditions. Both of these variables have the potential to cause an imbalance between power generation and load demand. This incompatibility may alter the system frequency to a critical degree, rendering the MG susceptible in terms of its reliability and security [6–8]. This necessitates the implementation of an efficient LFC approach that aims to maintain a balance between the power generation and load demand, prohibiting deviations in the system frequency within some predetermined standards, and ensuring the MG's operation is reliable and secure.

1.5 MODELING OF A MICROGRID AND FREQUENCY DEVIATION RESPONSE

In order to have a fair understanding of the FDR of an MG, it is vital to develop its frequency response model. The frequency response model is a simplified model based on neglecting nonlinearities but simultaneously it is believed to include the requisite system dynamics [9]. In this section, initially a general structure of an MG is presented and discussed. Later, the FDRs of a case study MG are examined.

1.5.1 GENERAL STRUCTURE OF AN MG

A generalized model of an MG is shown in Fig. 1.4. It consists of diverse micro sources that include n number of DGs, m number of energy storage units (ESUs), k number of electric vehicles (EV), a WTG unit, and a solar PV unit, all connected to an AC bus. The DGs may be a conventional diesel engine generator (DEG), a microturbine (MT), or a fuel cell (FC). The ESUs may comprise a battery energy storage unit (BESU), flywheel energy storage unit (FESU), redox flow battery unit (RFBU), or any other ESU. Considering a rapid development in the EV technology over the last two decades and its deployment in the power system for an enhanced frequency control [10–17], a fleet of the EVs is also incorporated in to the structure of the MG as shown in the figure. Each micro source is interfaced with the AC bus via a power electronic converter (PEC). For synchronization of the AC power generating units like the DEG or the WTG, AC/AC PECs are implemented while DC/AC PECs are utilized for synchronization of the DC power generating units like the solar PV and various ESUs. Furthermore, a circuit breaker (CB) is also installed between each of the micro sources and the AC bus to disconnect them from the utility grid in case of any severe disturbance in the grid or when the MG is scheduled for maintenance.

For portraying the dynamics of the various micro sources integrated in the MG, several linear models can be located in the quality literature [3, 6, 8, 15, 17–24]. All these models have been presented in detail in the Chapters 7, 8, and 9. To control the FDR of the MG, a feedback controller is enforced to regulate the output power of the various DGs and/or the ESUs [25, 26] and

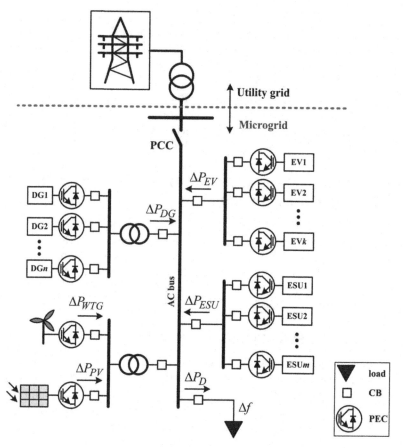

Figure 1.4   General structure of an MG

the WTG unit [13, 27] in response to the frequency deviation. Owing to the intermittent and unpredictable nature of the RES-dependent units like the solar PV and the WTG, the feedback controller must be capable of arresting the system frequency deviations in the least possible time failing which may result in unstable and undesirable MG operation.

1.5.2   FDRs OF AN MG: A CASE STUDY

Here the FDRs of an islanded MG are presented and analyzed. A linearized transfer function model is utilized for this purpose as shown in Fig. 1.5 [6]. The model comprises of a DEG, an FC, a solar PV, and a WTG as the DGs and a BESU, an FESU, and an EV model as the ESUs. A proportional integral (PI) controller is implemented as the feedback controller to regulate the output powers of the DEG, the FC, and the EV models. Relevant modeling details of the various DGs and ESUs incorporated in the MG are discussed in the upcoming chapters of this book. Effect of the PI controller on the FDRs is examined considering different combinations of the DGs and the ESUs.

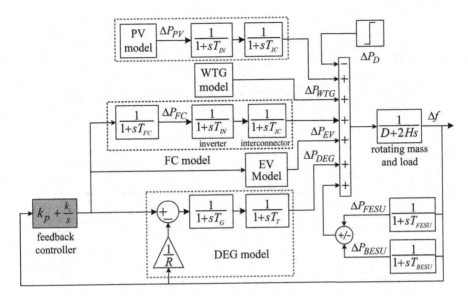

Figure 1.5   Linearized transfer function model of the case study MG

Figure 1.6 shows comparative FDRs of the MG considering 'without controller', 'with controller (DEG+EV)', 'with controller (DEG+FC)', and 'with controller (DEG+FC+EV)' cases subject a load perturbation of 0.01 per unit (pu). It is seen that for the 'without controller' case, the FDR attains a

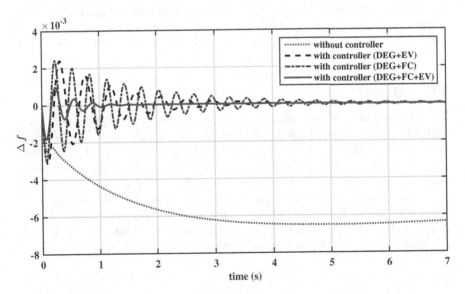

Figure 1.6   Comparative FDRs of the case study MG

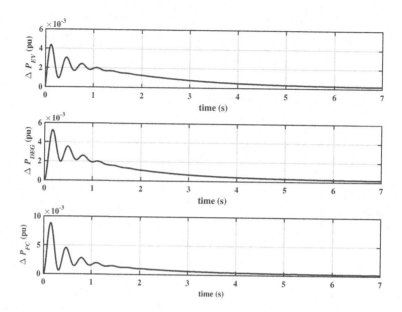

Figure 1.7   Power outputs of all the DGs controlled by the feedback controller

permanent offset from the reference value (i.e. 0). For the case 'with controller (DEG+FC+EV)', the MG exhibits a superior FDR as compared to the other cases. Power outputs (in pu) of the controlled units for this case are shown in Fig. 1.7. It can be seen that each of these units responds to the controller signal input to them in accordance with the frequency deviation (Δf) and restores the system frequency to normal in minimum possible time.

1.6   SUMMARY

This chapter provides an analysis of the several stability aspects associated with the power system. In addition to the traditional stability classifications, this discussion also highlights two recently established stability classifications, namely, resonance stability and converter-driven stability. The importance of the LFC in the power system is underscored, and an examination of several types of frequency control schemes is undertaken. The perspective of the MG about the LFC is depicted, and a generalized model of an MG is introduced and comprehensively discussed. Simultaneously, FDRs of a case study islanded MG are investigated incorporating a PI controller.

QUESTIONS

1. Discuss about the various power system stability issues taking into consideration the impact of the CIG technology.
2. What are the various control schemes implemented in the power system? Elaborate.
3. Give a detailed description of a generalized MG along with its various micro sources.
4. Investigate in detail about the two newly introduced power system stability types by the IEEE PES joint task force in the year 2016.

2 Renewable Energy Sources

This chapter begins by providing an introductory justification for the require-ment of the renewable energy sources (RESs) together with their related op-portunities involved in order to accomplish sustainable development. After that, a discussion of the benefits of the RESs is followed by an examination of the historical, contemporary, and prospective facets of the generation of electrical power from the RESs. The current status of the installed capacity of the RESs across India and the world is also comprehended. In conclusion, a brief discussion is presented on a variety of the RESs available.

2.1 RENEWABLES FOR SUSTAINABILITY

An RES is an energy source that either cannot be depleted or can be replen-ished on an ongoing basis. The stocks of fossil fuels are continuously dwindling, whereas the deposits of the RESs are endless and need to be exploited in an efficient manner. Although fossil fuels such as coal, oil, and natural gas have been the driving force behind the energy boom in recent centuries, at the same time, not only are their deposits being depleted at a petrifying rate but also emission of harmful pollutants from them is gradually degrading the environ-mental health. Although fossil fuels have been the driving force behind the energy boom in recent centuries, this does not change the fact that their de-posits are being depleted at a concerning rate. Figure 2.1 shows the emission of CO_2 during power generation from 1990 to 2019 [28]. A steady increase in the amount of CO_2 released into the atmosphere can be seen to be taking place during the course of this period in each successive year. In such a situa-tion, it would be necessary to make use of alternative energy sources that have the potential to be supplied for a very very long time and that function in an environmentally friendly manner. Several countries from all over the world have recently begun making investments in the research and development of various renewable energy technologies.

The consumption of energy is thought to be sustainable if it satisfies the current energy needs without compromising the requirements for energy that will be required by future generations. The world's growing population is driv-ing up the need for electricity, which in turn necessitates the exploration, de-velopment, and utilization of the RESs to facilitate the growth of the modern society. Although conventional fossil fuels still satisfy the majority of society's energy requirements, the drawbacks associated with these fuels, which were discussed earlier, as well as their increasing price, will limit their application in the future decades. As a result, investigating the RESs is essential at this point of time in order to achieve sustainable development. The development of the technology behind renewable sources of energy has seen a rapid ascent

DOI: 10.1201/9781003477136-2

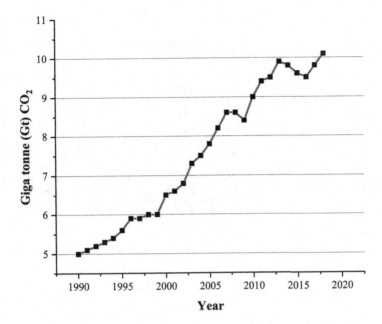

Figure 2.1 Global energy-related carbon dioxide emissions from 1990 to 2018 (Source: International Energy Agency (IEA) report 2019)

up until this point, but there are still a great many benchmarks to be reached. According to a report published by the International Energy Agency (IEA), it is anticipated that the amount of power generated by the RESs will increase by more than 60% between the years 2020 and 2026 [29].

2.1.1 OPPORTUNITIES OF THE RENEWABLES FOR SUSTAINABLE DEVELOPMENT

Figure 2.2 presents the various opportunities of the renewables for a sustainable development [30]. These include: (i) social and economic development, (ii) improvement in environmental and human health, (iii) energy access, and (iv) energy security. All these opportunities are briefly discussed below.

1. *Social and economic development*: Traditionally, the belief has been that the energy sector is essential to the expansion of the economy. This is due to the robust association that exists between expanding economic activity and rising energy consumption. Consumption of energy and per capita income are positively associated across the globe, and the current rise in energy consumption is mostly attributable to the development of economic activity. As a direct result, new jobs are created.

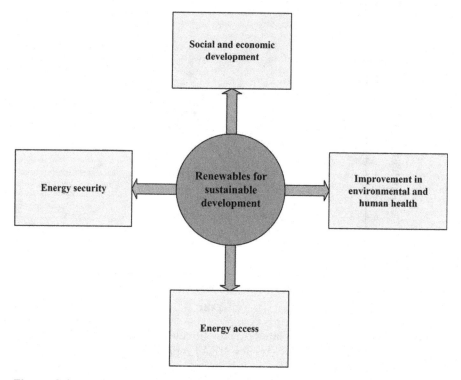

Figure 2.2 Opportunities of the RESs for sustainable development

2. *Improvement in environmental and human health*: When used in the generation of energy, the RESs contribute to a reduction in greenhouse gases, which in turn serves to delay the progression of climate change and the environmental and health concerns caused by pollution caused by conventional energy sources that are based on fossil fuels.

3. *Energy access*: In rural areas far from the utility grid, distributed networks based on the RESs are typically more competitive, and the low levels of rural electrification present considerable opportunities for RE-based (or RES-based) MG systems to supply them with electricity.

4. *Energy security*: When compared to fossil fuels, the RESs have more consistent distribution across the world and are generally sold on the market less frequently. The use of the RESs results in a number of potential benefits, including the reduction of energy imports, the creation of a portfolio of supply alternatives that is more diverse, the lessening of the economy's sensitivity to price volatility, and the potential for an improvement in global energy security. The installation of the RESs can help improve the dependability of energy services, in particular in locations that typically suffer from

inadequate grid connectivity. Utilizing a wide variety of energy sources, practicing effective management, and developing systems that are designed effectively are all ways to boost one's level of security.

2.1.2 ADVANTAGES OF THE RENEWABLE ENERGY

The RE retains numerous advantages few of which have already been highlighted. Here all the key advantages of the RE are presented along with some of their limitations [31].

Advantages of the RE include:

1. *Presence in abundance*: Since the RE is derived directly from naturally occurring resources (such as the sun, wind, and water, etc.) that are in plentiful supply, there is no possibility that the RE will ever become exhausted or depleted.

2. *Lower maintenance requirements*: When compared to conventional generators powered by fossil fuels, those powered by the RE require significantly less maintenance due to the presence of extremely few or no moving parts. This contributes directly to the reduction of costs in both time and money.

3. *Save money in long term*: Because the RE-based generators use the naturally occurring and abundantly available resources rather than the fossil fuels, this fact helps in saving money over the long run on operation, and it also reduces the amount of frequent maintenance that is required, which helps save even more money.

4. *Human health and environmental benefits*: Conventional power plants that run on fossil fuels release pollutants that are hazardous to people's health as well as to the health of the environment. On the other hand, the RE-based generators produce almost no harmful emissions, and as a result, they are beneficial to the health of both humans and the environment.

5. *Lower reliance on imported energy*: Since the RE-based generators can be installed locally to generate power to feed load, this directly reduces the reliance on the imported energy.

In spite of the several advantages related with the RE, it possesses some limitations also. These are:

1. *Higher initial cost*: Although the RE-based generators have very less running and maintenance costs, the utilization of expensive technology for power generation makes their initial cost comparatively much higher.

2. *Intermittency of the RE*: Compared to the power generation from the fossil fuel-based generators that is continuously available, the RE-based generators rely on weather conditions that are not the same the whole day. Hence, the power output becomes intermittent.

3. *Need for energy storage systems*: Owing to the intermittency of the RE, it becomes mandatory to utilize energy storage systems to maintain the continuity of supply of the power generated. The energy storage systems are quite expensive specifically for large-scale power generation.

4. *Geographic limitations*: An RE-based generator can be installed only at specific locations in accordance with the availability of that particular RE. For example, wind farms can only be installed in large open areas where wind speed is considerable the whole day.

2.1.3   STATUS OF THE RESs: INDIA AND THE WORLD

The RESs are currently regarded as a potential answer to address the energy challenges of the future. Currently, governments worldwide are engaged in the process of developing policies, programs, and strategies aimed at attracting international investments in research and development of RES technologies for domestic use. The imperative for utilizing the RESs has become increasingly evident, as they play a crucial role in bolstering energy security and efficiency, fostering an economy characterized by reduced carbon emissions, and facilitating the expansion of energy accessibility. Nevertheless, the reliance of the RESs on weather conditions is a significant obstacle to their seamless integration into the primary power system, necessitating effective management and control measures. This section presents an analysis of the current state of the RESs for power generation in both India and globally. Additionally, it offers an overview of the present status of the RESs and their anticipated future trajectory.

2.1.3.1   Current status and future trend of the RESs in India

India currently has an installed capacity of 424.267 GW as reported on 31.08.2023 by the Central Electricity Authority (CEA), Government of India [32]. This is shown bifurcated in Fig. 2.3. Out of this total installed capacity thermal (coal + lignite + gas + diesel) accounts for 238.442 GW sharing 56.20% of the total, nuclear account for 7.48 GW sharing 1.76% of the total and RESs (including hydro) account for 178.364 GW sharing 42.04% of the total. Gradually expanding Indian economy, due to increasing urbanization and industrialization, is a strong indicator that energy demand will increase in the coming years. Owing to the ill effects related to the fossil fuel-based sources, it will be of prime importance for India to shift to the RESs as much as possible to meet the growing energy demand.

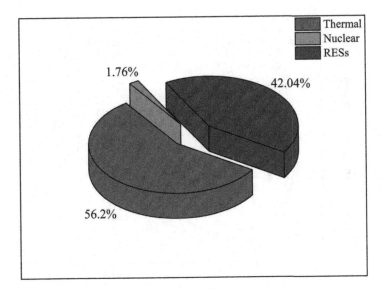

Figure 2.3   Installed capacity as on 31.08.2023 (Source: Central Electricity Authority (CEA), Government of India)

Figure 2.4 shows the forecast electricity demand for India till the year 2040 which is around 4000 terra watt hour (TWh) [33]. Figure 2.5 shows the

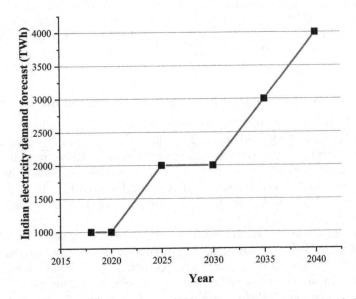

Figure 2.4   Forecast of the electricity demand in India (2018–2040) (Source: World Energy Outlook 2019, International Energy Agency (IEA))

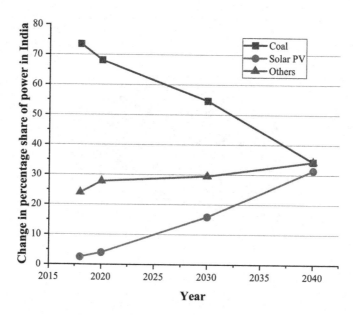

Figure 2.5 Change in percentage share of power generation in India (2018-2040) (Source: India Energy Outlook 2021, International Energy Agency (IEA))

changes in the percentage share of power generation from coal, solar PV, and other sources till the year 2040 [34]. It is estimated that the percentage share of generation from coal would decrease from 73.5% in 2018 to 34.3% in 2040, whereas for the solar PV the estimate in the percentage share would increase from 2.5% in 2018 to 31.4% in 2040.

2.1.3.2   Current status and future trend of the RESs in the World

As per the IEA report 2022, the total power generation from the RESs world-wide in the year 2021 was 7964.8 TWh [35]. Figure 2.6 shows a bifurcated pie-chart showing the percentage share of the various RESs, namely, hydro, wind, solar PV, bioenergy, and others. It can be observed that power generation from hydro has a share of above 50% followed by wind, solar PV, bioenergy, and others. As per the report, the share of the RESs for electricity generation was 28.7% in 2021.

Figure 2.7 shows the worldwide electricity demand forecast (TWh) for various sectors till the year 2040 [33]. It can be seen that the energy demand is continually dominated by the industry, residential, and services sectors. Energy demand of the transport sector can be seen increasing after the year 2030. It will be due to the emergence of the EVs. Figure 2.8 shows the electricity generation forecast till the year 2040 [33]. It is seen that the generation

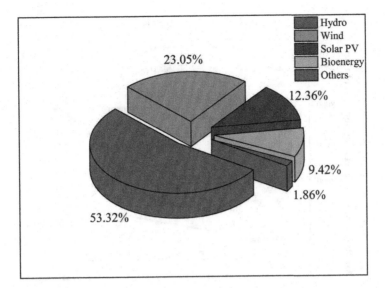

Figure 2.6 Power generation from the RESs in 2021 (Source: International Energy Agency (IEA) report 2022)

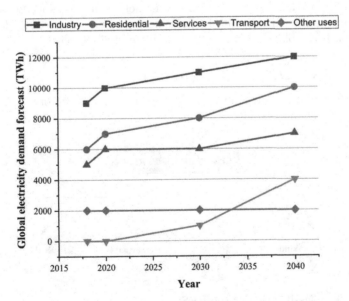

Figure 2.7 Global electricity demand forecast (2018–2040) (Source: World Energy Outlook 2019, International Energy Agency (IEA))

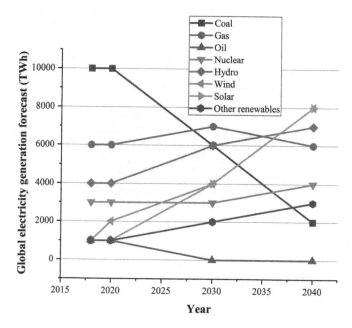

Figure 2.8 Global electricity generation forecast (2018–2040) (Source: World Energy Outlook 2019, International Energy Agency (IEA))

from the conventional sources of energy namely coal, gas, and oil would be discouraged gradually till 2040. It is also noteworthy that craze for electricity generation from solar and wind would be increasing the most.

2.2 CLASSIFICATION OF THE RESs

Figure 2.9 presents the classification of the RESs. These are briefly discussed in the following subsections.

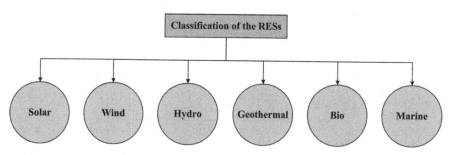

Figure 2.9 Classification of the RESs

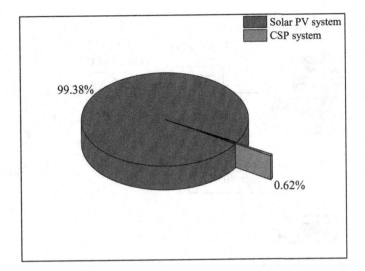

Figure 2.10 Percentage share of the total installed solar power generation capacity worldwide between solar PV and CSP systems for the year 2022(Source: Renewable Capacity Statistics 2023, IRENA)

2.2.1 SOLAR ENERGY

Solar energy is being utilized all around the world for a variety of applications, including the generation of electricity, the heating of water, and the removal of salt from the water. The generation of electricity is by far the most important use of solar energy. According to the most recent data published by IRENA [36], the global installed capacity for solar energy increased by close to 192 GW in the year 2022. Both China (with over 86 GW) and the United States (with nearly 18 GW) were significant contributors. About 13 GW worth of additional capacity was installed in India. The sun rays can be used in one of two ways to generate electricity: either through a solar photovoltaic PV system or through a concentrated solar power (CSP) system. Both of these methods can be utilized. The solar PV system is the one that sees the most use of the two options. Figure 2.10 depicts, for the year 2022, the percentage of installed capacity of solar photovoltaic systems and concentrated solar power systems over the world.

2.2.1.1 Solar PV system

A solar PV system consists of several PV cells connected in series or parallel. A PV cell is a solid-state electrical device that directly converts the photon energy into electricity. Basically the PV cells can be of three types:

1. crystalline silicon cell
2. amorphous silicon thin film cell

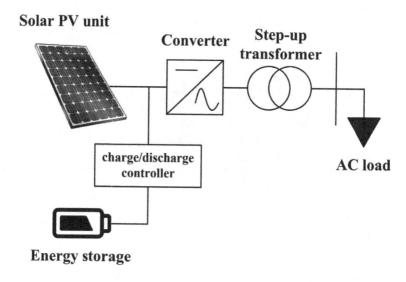

Figure 2.11 Scheme of an independent solar PV system

3. hybrid thin film cell

The crystalline silicon cell is the type of solar PV cell that is used in the solar PV system the most frequently. A solar PV system has many benefits, some of which are as follows: (i) it does not pollute the environment and its operation is completely silent, (ii) it lowers the cost of monthly electricity bills since there is an abundance of sunlight, (iii) it has a long life and requires very little maintenance, and (iv) it is simple to install either on a rooftop or on the ground. Concurrently, the solar PV system has a number of drawbacks, including the following: (i) it is less reliable due to the intermittent and unpredictability of the solar energy, (ii) it has a low operational efficiency of between 14% and 25%, and (iii) it has a greater installation cost.

It is possible for a solar PV system to function either independently or while connected in parallel to the utility grid [37]. In order to meet the requirements of the local load demand, an independent solar PV system is typically installed in a remote location. The solar PV system ought to have the ESUs as a backup in order to make up for the fact that the sun does not provide energy continuously throughout the day and that the load demand has to be satisfied on a consistent basis. The ESUs have a tendency to discharge to feed the load when the solar energy is not available, while they charge when it is available. Figure 2.11 illustrates the fundamental layout of an independent solar PV system that can function on its own. In contrast, a solar PV system that is connected to the utility grid typically sells the electricity that it generates to the utility. The fundamental layout of a solar PV system that is connected to and operates in conjunction with a utility grid is depicted in Fig. 2.12.

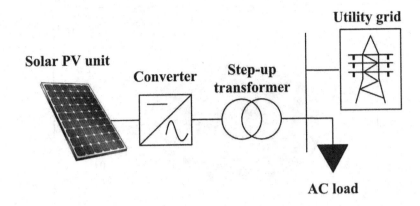

Figure 2.12 Scheme of a solar PV system connected to the utility grid

In addition, the solar PV system that is connected to the utility grid can be subdivided into two different types: the distributed type and the centralized type. The distributed type solar PV system supplies the loads with power directly, and the utility grid is responsible for making up for any shortfall or surplus in power supply. A solar PV system of the centralized type, rather than serving the loads directly, injects the electricity into the utility grid so that it can be distributed further.

2.2.2 WIND ENERGY

Through the use of a turbine and a generator, wind energy can be converted into usable forms of energy, including electricity. The generator takes the mechanical energy that is produced by the turbine and transforms it into electrical energy. This process begins when the kinetic energy of the wind is converted into the mechanical energy of the turbine. Wind power is quickly becoming one of the most popular RESs all over the world. According to the most recent information from IRENA [36], the installed capacity for wind energy (both onshore and offshore) has expanded by a factor of three since the beginning of the previous decade, going from a mere 300.04 GW in the year 2013 to 898.82 GW in the year 2022. Between the years 2021 and 2022, a total of 74.65 GW of additional installed capacity was produced, which is approximately equivalent to an increase of 9%.

Onshore wind and offshore wind each make up their own portion of the total installed capacity for generating electricity from the wind in the year 2022, as shown in Fig. 2.13. The wind farm that is positioned on land is referred to as an onshore wind farm. The capacity of the wind turbines in these farms is approximately 2 MW. The term offshore wind farm refers to a wind farm that is constructed in vast bodies of water, which often have a higher average wind speed. The capacity of these wind turbines ranges between 3 and 5 MW.

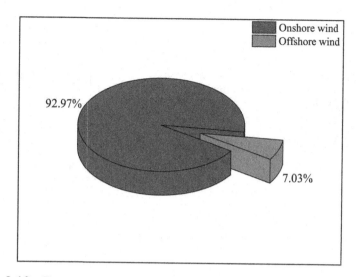

Figure 2.13 Percentage share of the total installed wind power generation capacity worldwide between onshore wind and offshore wind for the year 2022 (Source: Renewable Capacity Statistics 2023, IRENA)

WTGs provide a number of benefits, including the following: (i) they do not contribute to environmental pollution, (ii) they have a long operational life, (iii) they have cheap operating and maintenance expenses, and (v) they do not prevent the land from being used for other purposes. In the same breath, a WTG has a few drawbacks, which include the following: (i) due to the intermittent and unpredictable nature of wind, energy storage is required, (ii) practicable only in locations with high wind speed, and (iii) extensive acreage is required for installation due to the low energy density of wind. Despite these drawbacks, the WTGs continue to be widely used as a source of renewable energy.

Like a solar PV system, a WTG system can also be either operated independently or in parallel connection with the utility grid [37]. Furthermore, when connected to the utility grid, a WTG system may be either distributed type or centralized type. The general scheme of an independent WTG system is shown in Fig. 2.14.

2.2.3 HYDRO ENERGY

The generation of electricity from moving water by means of a turbine and generator set is referred to as hydro power. This type of energy is also known as hydro energy. The moving water is channeled to fall from a significant height onto a mobile turbine, which then rotates a generator that is coupled to it. Electricity is produced as a result of the generator's rotation. The most efficient and economical method of producing electricity is through the use of

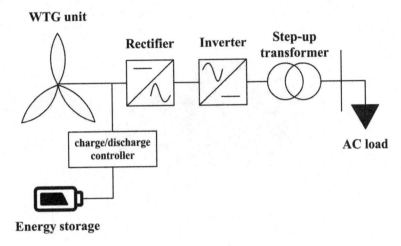

Figure 2.14 Scheme of an independent WTG system

a hydroelectric power plant, which may be broken down into three distinct types: (i) impoundment type (with reservoir), (ii) run-of-river type (without reservoir), and (iii) pumped storage type [38]. The most common type of hydro power plant is an impoundment type plant, which stores water in a big reservoir. These plants generate hydroelectricity. After the water has been held, it is released so that it can flow over a turbine-generator set to produce power. A run-of-river type plant does not have any reservoirs; rather, it causes the water of any river to be diverted and flow via a channel or penstock. After being redirected, the flow of water is utilized in the production of electricity. During non-peak hours, the water in a plant of the pumped storage type is transferred by pumping to a reservoir at a higher elevation. During the peak hours of the day, the water that has been pumped up is directed to flow into a lower reservoir, which then rotates the turbine generator set, resulting in the production of energy. Electricity generation in remote or isolated places might benefit from the use of micro-hydro power plants operating on a smaller scale.

As per the latest report of the IRENA [36], total global installed hydro power generation capacity in the year 2022 was nearly 1392 GW. As compared to the year 2013, when the installed capacity was 1137.292 GW, the overall increase is approximately 18%. Figure 2.15 shows the bifurcation of the installed hydro power generation capacity for the year 2022. However, percentage increase in the installed capacity of the hydro power generation has been less compared to the installed capacities of the solar and wind energies over the past decade but still it continues to be the RES with the highest installed capacity.

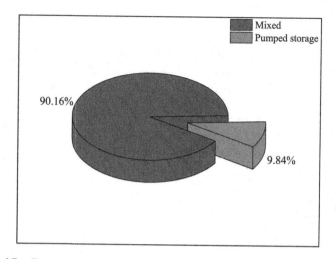

Figure 2.15 Percentage share of the total installed hydro power generation capacity worldwide between mixed (impoundment type + run-of-river type) and pumped storage type for the year 2022 (Source: Renewable Capacity Statistics 2023, IRENA)

2.2.4 GEOTHERMAL ENERGY

The release of geothermal energy from the earth's crust is a result of the decay of radioactive elements. This energy is subsequently transferred to the subsurface through the processes of conduction and convection. This is the process by which geothermal energy is generated. Geothermal energy possesses the capacity to facilitate power generation, with optimal resources for this purpose being characterized by elevated or moderate temperatures and close proximity to places exhibiting tectonic activity. Furthermore, geothermal energy can be utilized in many applications such as geothermal heat pumps, which are employed for the purpose of heating and cooling buildings, as well as for space heating and bathing. Geothermal power production technology has achieved significant advancements and is being employed in around 26 countries worldwide [39].

Dry steam power plant, flash steam power plant, and binary cycle power plant represent several technological advancements that are currently employed in power generation facilities. In a dry steam power plant, if steam is the primary fluid collected, it is used directly to operate turbines connected to generators. The steam's pressure and temperature are then utilized to generate mechanical effort, which generates electricity. When bringing hot water to the surface, it is transported through a separator or flash tank. The abrupt decrease in pressure causes the water to be converted into steam, which is then used to drive turbines and generate electricity in case of a flash steam power plant. In situations where the temperature of the geothermal resource is insufficient to produce steam directly for power generation, a binary cycle power plant is utilized. The geothermal fluid's heat is transferred to a secondary fluid (such as isobutane or isopentane) with a lower boiling point, which vaporizes and drives a turbine connected to a generator.

The primary advantage of utilizing geothermal energy is in its immunity to meteorological conditions, enabling its utilization at any given moment, regardless of day or night. As a result of this, geothermal power plants have the capability to simultaneously fulfill essential energy demands and offer additional support services. According to the latest report, the global installed capacity for geothermal electricity generation in 2022 amounted to 14,877 MW. According to a report by IRENA [36], there has been a significant increase in capacity, with a growth rate of around 35.45% observed over the past decade.

2.2.5 BIOENERGY

The energy derived from the process of combusting biofuels is commonly known as bioenergy or biofuel energy. Biofuels are defined as the biomass of plants that possess the capacity to store energy in the form of chemical energy. Upon combustion, the stored chemical energy is liberated and transformed into thermal energy. The study conducted by [40] reveals that biofuels have the potential to manifest in several physical states, including solid, liquid, or gaseous. The constituents comprising solid biofuels encompass animal waste, charcoal, wood chips, pellets, and firewood. Solid biofuels have been utilized by people for a considerable duration in the past for the purpose of generating heat and light, spanning a significant portion of human history. Currently, there exist three distinct categories of liquid biofuels, namely, bioethanol, biodiesel, and pyrolysis bio-oil. The transportation business predominantly utilizes liquid biofuels for various applications. Gaseous biofuels consist of two primary constituents, namely, syngas and biogas. The creation of syngas is a result of the gasification or pyrolysis process applied to plant matter. The 2023 report of IRENA suggests that developing economies, including Brazil, India, and China, exhibit substantial prospects for bioenergy-based electricity generation [36]. This potential is expected to effectively address the escalating power requirements of these nations. Based on the latest data provided by the IRENA report [36], the aggregate installed capacity for bioenergy in the year 2022 amounted to 148.912 GW. In Fig. 2.16, a comprehensive depiction is provided for the given year, illustrating the categorization of various biofuels according to their respective states, namely, solid, liquid, and gaseous. The image also portrays municipal waste that is capable of being recycled, commonly known as garbage or municipal solid waste.

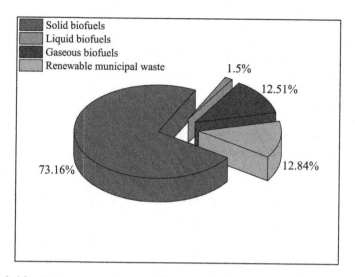

Figure 2.16 Percentage share of the total installed bioenergy power generation capacity worldwide between solid, liquid, gaseous and municipal waste biofuels for the year 2022 (Source: Renewable Capacity Statistics 2023, IRENA)

2.2.6 MARINE ENERGY

The term marine energy is commonly used to describe the energy harnessed from ocean waves, tides, and currents, while it is occasionally referred to as ocean energy. The kinetic energy of these things can be harnessed and transformed into electrical energy, serving a multitude of applications. Despite being in its nascent phases of development and lacking maturity, maritime energy technology is anticipated to have substantial promise in the realm of power generation. According to [41], there exist three distinct categories of marine energy technology that can be harnessed for power generation: (i) wave energy, (ii) tidal energy, and (iii) salinity gradient energy. The cumulative installed capacity of marine energy has achieved a significant milestone of 524 MW as of the year 2020 [42].

2.3 CHALLENGES AND FUTURE TRENDS

This section presents a comprehensive discussion on the challenges related to the RES-based generators followed by the future trends in their implementation.

As per [43], the solar PV systems and WTG systems have been the most widely implemented RES-based generators in the preceding decades. Their dependence on weather conditions results in a fluctuating power output from them. Consequently, following challenges are presented during their integration:

1. *Power quality*: The integration of the RES-based generators has been found to have several negative effects on the power quality. These include an increase in flickers, higher levels of harmonic distortions, unreliable shut downs, and more variations in local voltage. As a result, the overall power quality is significantly degraded. Various causes have the propensity to cause harm to the equipment and result in a reduction of their operational lifespan.

2. *Power flow*: The RES-based generators exhibit irregular power output, leading to elevated local voltage levels, significant short-circuit currents during faults, insufficient protection mechanisms on the low-voltage side, and inadequate transmission capacity. The combination of these elements results in unanticipated power flows on the low-voltage side, posing a risk to equipment functionality and diminishing their operational lifespan.

3. *System stability*: When comparing the RES-based generators to the fossil fuel-based generators, it is observed that the RES-based generators exhibit a lower reactive power output, possess a lower degree of short-circuit power, and demonstrate an overall decrease in rotational inertia. The aforementioned factors contribute to issues of instability in voltage and frequency on the low voltage side. Moreover, the lack of proper coordination between voltage and frequency trip limits leads to issues of dynamic stability and undetected oscillations in power flow.

4. *Power balance*: The incorporation of RES-based generators introduces fluctuating performance demands for fossil fuel-based generators, thereby giving rise to both immediate and prolonged challenges in power generation. These issues aggravate the discrepancy between power generation and load demand. Moreover, the discrepancies in the prediction accuracy pertaining to the RES-based generators contribute to the occurrence of generation-load mismatch.

According to the Global Energy Transformation report published by IRENA in 2018 [44], it is projected that the RES will account for 85% of the total electricity generated in the year 2050, a significant increase from the 24% recorded in 2015. By the year 2050, wind energy is projected to become the predominant RES for power generation, with solar energy following closely behind in terms of widespread utilization. The projected global installed capacity for different RESs in the year 2050 is presented in Table 2.1. Moreover, according to the research, there is a projected reduction of approximately 85% in CO_2 emissions by the year 2050 in comparison with the year 2015. It is shown as well in the table. Furthermore, it is projected that the utilization

Table 2.1

Forecast in the installed capacity of the RESs in 2050

RES	Installed capacity (GW) in 2015	Installed capacity (GW) in 2050
Hydropower	1248	1828
Wind	411	5445
Solar PV	223	7122
CSP	5	633
Bioenergy	119	384
Geothermal	10	227
Others (marine, etc.)	0.3	881
CO_2 emission (Gt/yr)	12.4	1.9

of the RESs in the transportation industry will increase to approximately 58% by 2050, in contrast to a mere 4% in 2015. According to the report, a projected outcome is the generation of about 11 million employment opportunities within the energy sector by the year 2050 as a result of the shift from fossil fuels to the RESs.

2.4 SUMMARY

This chapter underscores the importance of the renewables in the pursuit of sustainable development, alongside the diverse array of opportunities associated with this attempt. Subsequently, a comprehensive examination of the current state of the RESs in India and globally is provided, which is thereafter followed by an analysis of the foremost advantages linked to the RE. Afterward, an analysis is conducted on several types of RESs, together with their respective latest global installed capacities, in a sequential manner. In conclusion, this chapter has examined the primary challenges associated with the integration of RES-based generators, followed by an overview of future developments in the utilization of different RES technologies.

QUESTIONS

1. Renewable energy is essential for sustainability. Elaborate the statement.
2. What are the various opportunities related to the renewable energy for sustainable development? Discuss.
3. Discuss the current status and future trends in the RESs in India as well as the world.
4. Discuss in detail the challenges related to the implementation of the RESs for power generation. Further, highlight the future trends of the RESs.

3 Distributed Generation and Energy Storage

The traditional power plants have been significantly compromising the environmental well-being of the global ecosystem through the production of harmful pollutants during the past few decades. Concurrently, the rapid depletion of fossil fuel reserves has emerged as a significant cause for apprehension. In order to address these challenges, it has become common practice to generate electricity on-site at the distribution level, specifically at medium voltage or low voltage. The word used to describe this form of electricity generation is known as the distributed generation (DG). Given the intermittent and unpredictable nature of specific DG sources, such as solar PV and WTG units, it is imperative to incorporate energy storage (ES) technology. The ES technology is specifically designed to enhance the operational performance of an MG. This chapter provides a comprehensive analysis of several DG sources and their associated advantages. Subsequently, the significance of the ESUs in the MG is underscored, leading to an extensive examination of the diverse ES technologies currently accessible.

3.1 WHAT IS DISTRIBUTED GENERATION?

As per the IEEE Industry Applications Magazine, DG is defined as [45]:

> *The power generation by the power generating units typically not larger than 1 or 2 MW that are installed mostly by the utility itself or end users and are connected to the utility distribution system.*

The principal reason behind the emergence of the DGs is the concerns over the environmental pollution from the fossil fuel based-generators and simultaneously, the rapid depletion of the fossil fuel deposits. The term 'distributed generation' has been proposed to distinguish it from the centralized conventional power generation.

3.1.1 BENEFITS OF THE DGs

Key benefits associated with the DGs are [46–50]:

1. *From economical viewpoint:*
    i. The DGs can easily feed the increased load demands by installing them in their vicinity and thus reducing the need of additional T&D

lines, losses, and power system expansion.

ii. The DG modules (combination of two or more DG sources) can be installed in a very short time span and operated independently. Simultaneously, the total installed capacity can be increased/decreased by varying the number of DG modules.

iii. Since the DGs can be installed at any desired location independent of the centralized power generation, the electricity prices are greatly affected.

iv. Size of the DGs is quite appropriate to be changed in small amounts to exactly match the required load demand of the end users.

v. The DGs can reduce wholesale power price by supplying power to the utility grid, thus, required demand is reduced.

vi. The DGs increase the lifetime of transformers and other system equipment and reduce fuel consumption.

vii. Diversification in the DGs reduces the need for certain types of fuels more than the others.

2. *From operational viewpoint:*

i. The DGs reduce the losses in the distribution network, reduce load demand on the distribution side by feeding a fraction of the connected load, reduce burden on the transmission lines, and improve their voltage profile.

ii. The DGs can facilitate peak load shaving and load management programs.

iii. Reliability of the system can be enhanced and continuity of power supply can supported with the DGs.

iv. The DGs improve the system stability and supply the spinning reserves when required.

v. The RES-based DGs are free from any harmful emissions and thus cater to improve the environmental health.

vi. Availability of the DGs in various power output capacities provide flexibility for sizing and siting of these in the distribution network.

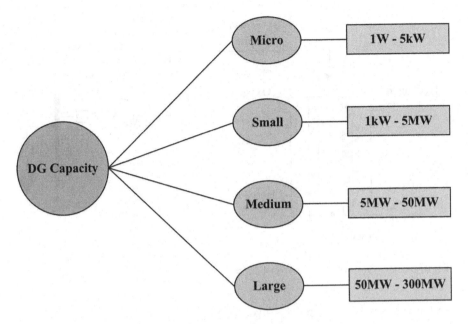

Figure 3.1   Classification of the DG capacities

> vii. The DGs can be put on a stand-by to supply power in case of any
> emergency or system outages.

3.1.2   DG CAPACITIES

Although not strictly specific, a broad classification of the DG capacities is
given in Fig. 3.1 [46]. The capacity can further be increased by combining two
or more DG units.

3.2   TYPES OF DG SOURCES

This section provides details about the various conventional DG sources implemented in an MG. These are fuel cell (FC), micro turbine (MT), and diesel
engine generator (DEG) [37,51–53]. The RES-based DG sources, namely, solar PV and WTG have already been discussed in the preceding chapter and
are presented here only in brief.

3.2.1   FUEL CELL

Electrochemical cells that convert the chemical energy of a fuel (often hydrogen) or any other fuel and an oxidizing agent (typically oxygen) into electrical
energy are referred to as fuel cells. An FC is a type of an electrochemical cell.
The FC is able to maintain its capacity to create electrical energy constantly

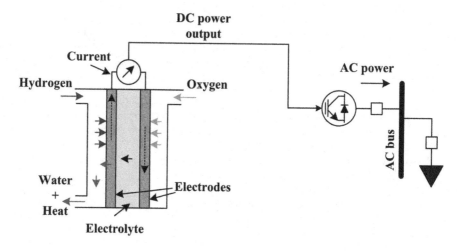

Figure 3.2   General structure of an FC

so long as a sufficient supply of hydrogen and oxygen is maintained. In 1939, a British engineer named Francis Thomas Bacon conceived of the idea for the first hydrogen-oxygen fuel cell, which had the capacity to generate a power output of 5 kW. Since that time, the FCs have found use in a wide variety of applications, such as space travel vehicles, automobiles, electronic devices, auxiliary power production, boats, and submarines.

General structure of an FC is made up of two porous electrodes (the anode and the cathode), with an electrolyte sandwiched in between them. Within the cell, there are separate pathways that are designed for the delivery of the fuel and the oxidant, respectively. Both of these gases are taken in by the porous electrodes, which then trigger a chemical reaction that results in the generation of electrical energy. The hydrogen-oxygen FC shown in Fig. 3.2 is the one that sees the most usage. In this part of the process, hydrogen and oxygen gases are forced to move through the electrodes, which are referred to as the anode and cathode, respectively. The resulting current is utilized, while the by-products, which consist of water and heat, are discharged through a different path, as illustrated in the picture. Because the FC produces a DC signal, it must first be converted into AC by an appropriate PEC before being sent to a load that is linked to an AC bus. Depending on the specific requirements of an application, an FC has the potential to produce a power output that can range anywhere from a few kW to several MW.

Currently, the FC is available in various types as shown in Fig. 3.3. The various fuels and electrolytes used and corresponding efficiencies are also shown in the figure. Principal advantages of an FC are:

(i) reduced greenhouse gas emissions
(ii) high efficiency

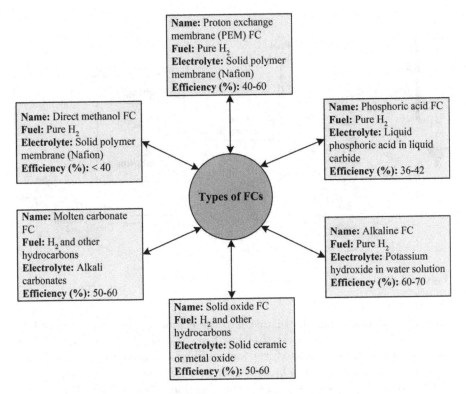

Figure 3.3 Classification of FCs

(iii) flexibility in installation and operation
(iv) low number of moving parts reduce, operating and maintenance costs
(v) increased reliability and
(vi) less dependence on foreign oil

3.2.2 MICRO TURBINE

An MT is a small gas turbine-based power generation system that can be operated on different fuels such as natural gas, propane, diesel, and light oil. An MT generally utilizes compressed air and a fuel that are burnt under constant pressure conditions to generate combined heat and power (CHP). Currently, MTs are available in power output ranging from 20 kW up to 500 kW and efficiency around 30% (without heat recovery) and around 85% (with heat recovery).

Figure 3.4 provides an illustration of the general layout of an MT. Compressor, combustor, turbine, generator, and heat exchanger are the primary components that make up this device. In the beginning, air is forced into the compressor, where both its temperature and pressure are raised to a higher

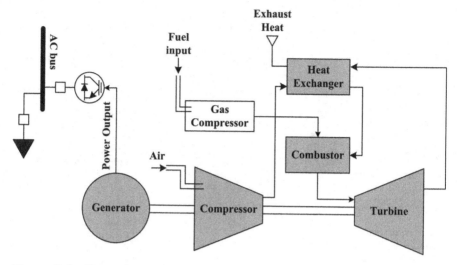

Figure 3.4   General structure of an MT

level. At the same time, fuel is fed into the combustor, where it comes into contact with high-pressure air that has been injected from the compressor. After being ignited and heated, the fuel-and-high-pressure-air mixture begins to progressively expand as it burns off the fuel. After that, the heated and expanded mixture is sucked into the turbine section, where it causes the blades of the turbine to rotate, and consequently the alternating current generator that is attached to the turbine shaft. As soon as the generator has reached a steady speed, it will begin producing electrical power. This power will then be supplied to the load that is linked to an AC bus by going via the appropriate PEC. The primary goal of utilising the heat exchanger is to improve the MT's overall operational efficiency. It does this by preheating the air that is going to be pumped into the combustor, which in turn lowers the amount of fuel that is used and, as a result, raises the overall efficiency of the system. Depending on the method that is used to generate power, an MT can be divided into one of two categories. The first type is the unrecuperated type, and the second type is the recuperated type.

Since an MT has the ability to generate both the active and reactive powers, it can control both the system frequency and voltage, thus ensuring a reliable and stable operation of an MG especially in the islanded mode. Key advantages of an MT are

(i) increased reliability of power supply due to its ability to operate in stand-alone mode
(ii) reduced greenhouse gas emission with use of natural gas as fuel
(iii) cost savings in energy due to reduced fuel usage
(iv) reduction in T&D losses due to installation on the end-user side

(v) cost-effectiveness due to high thermal efficiency
(vi) low noise during operation and
(vii) much longer life than a DEG

3.2.3 DIESEL ENGINE GENERATOR

A DEG uses diesel or biodiesel as the input fuel for electric power generation. The energy from the fuel is changed into mechanical energy through the use of an internal combustion engine, and then the mechanical energy is changed into electrical energy through the utilization of an electric generator. In broad terms, a DEG is a member of the same family as other types of internal combustion engines. Depending on the capacity of its tank and the amount of fuel that is currently contained within it, the ability of a DEG to make the electricity it generates accessible at any location and at any time is possible. Continuous load tracking is carried out by the DEG, which also has the capacity to compensate for any load variation or supply disruption that may occur within the system.

As far as the operating characteristics of the DEG are concerned, it has the ability to provide a quick and reliable power supply during the peak load hours, has a rapid response time (within 10 s), is highly durable (may operate from tens to thousands of hours before its first overhaul), and is easy to install and operate. During the normal power system operation, the DEG is often kept as a standby power source to avoid extra cost on energy tariff and is operated only to supply the load during peak hours or when the main generators are unavailable for power generation. It can also be implemented to feed important stand-alone applications in case of any emergency. When connected in an islanded MG, the DEG engages in load tracking and provides quick and reliable power supply.

General structure of a DEG feeding a load connected to an AC bus is shown in Fig. 3.5. Initially, air is fed to the internal combustion engine until it

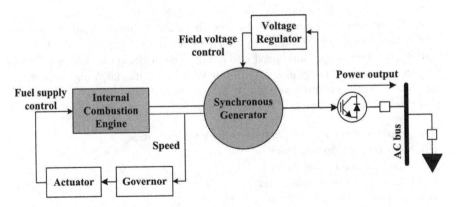

Figure 3.5   General structure of a DEG

gets compressed. Afterward, a fuel (generally diesel or biodiesel) is injected in the engine that comes in contact with the compressed air and produces heat. Subsequently, this heat triggers the combustion of the fuel that results in the starting of the synchronous generator. Finally, the synchronous generator generates electrical power as output that is supplied to the AC bus via a PEC and fed to the load connected to the bus. It is noteworthy here that a DEG can simultaneously inject both active and reactive powers to the bus. Potential advantages of a DEG include:

(i)   relatively low maintenance required
(ii) highly durable to changing environmental conditions
(iii) more rugged and reliable
(iv) relies on diesel for operation which is easily available fuel anywhere and
(v)   longer runtime

3.2.4 WIND TURBINE GENERATOR

Wind energy is available throughout the year in the majority of the global locations. Because of this, it is regarded as one of the most reliable methods for producing electrical energy. As has already been dealt in Chapter 2 of this book that a WTG refers to the equipment utilized in the conversion of wind energy to electrical energy. Wind energy is used to power a turbine consisting of a rotor with two or three affixed blades. The primary shaft, which is connected to the turbine, turns the generator that is utilized in the process of producing electrical energy which is then supplied to the utility grid. Figure 3.6 shows the schematic of a WTG.

A WTG exclusively relies upon the wind for its operation. The wind turns the rotor blades, and hence, the generator starts rotating and generates electrical energy. There are two categories of the WTG: horizontal-axis and vertical-axis. Typically, horizontal-axis WTGs have three blades and operate in the wind. The blades of this form of turbine pivot at the top of the tower to face the wind. The vertical-axis WTGs are omnidirectional. This indicates that they do not need to be adjusted to face the wind in order to function. Generally, horizontal-axis WTGs are preferred over the vertical-axis WTGs. It is noteworthy here that wind speed and swept area of the rotor blades influence the amount of electricity generated by a WTG. Additionally, air density and the power coefficient influence the WTG power output.

Principal advantages of a WTG unit are:

(i)   simple and only periodically required maintenance
(ii) good conversion efficiency
(iii) little environmental impact
(iv) limited land occupancy and
(v)   low installation and running costs

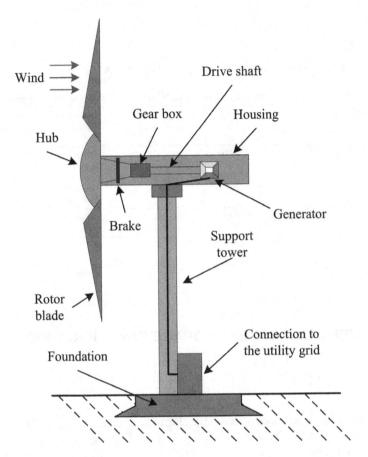

Figure 3.6   Schematic of a WTG

3.2.5   SOLAR PV MODULE

Chapter 2 of this book already provides an extensive analysis and examination of the solar PV system. Therefore, only a limited number of exclusive facts regarding it are being offered in this context. Due to the fact that solar energy is accessible in nearly all locations across the globe, the PV modules have emerged as the most viable option for meeting the demand for electricity in remote regions and in the MGs. The amount of solar irradiation, total area of the module, and its efficiency all have an effect on the amount of electrical energy that is generated by a PV module. It is noteworthy here that a solar PV module is only capable of generating active power.

A comparative summary of the various DG sources in terms of capacity range, efficiency, fuel used, and major applications is given in Table 3.1.

Table 3.1

Comparative summary of the various DG sources

DG source	Capacity range	Efficiency (%)	Fuel used	Applications
Fuel cell	1 kW – 20 MW	30 – 60	Natural gas, biogas, propane	Transportation, backup power
Micro turbine	25 kW – 500 kW	20 – 30	Natural gas, biogas, propane, diesel, hydrogen	Cogeneration, backup power
Diesel engine generator	20 kW – 480 kW	30 – 35	Liquid fuel, natural gas	Transportation, drilling
Wind turbine generator	300 kW – 5 MW	20 – 40	Wind	Agriculture, residential
Solar PV cell	300 kW – 2 MW	5 – 15	Sunlight	Transportation, power in space

3.3   NEED OF ENERGY STORAGE IN A MICROGRID

Energy storage technology is an integral part of an MG that influences its reliability and stability. Key reasons why the ESs are needed are summarized below:

1. *Renewable energy integration:* By charging and discharging, the ESs make up for the intermittent power outputs of the RES-based generators (such as solar PV and WTG) integrated in the MG and guarantee an uninterrupted power supply to the end users.

2. *Load balancing:* Due to their fast response time, the ESs assist the various DG sources to counterbalance any abrupt load deviations in the MG and thus, sustaining a balance between the generation and load.

3. *Support to the utility grid:* In certain instances, MGs are capable of providing grid support services to the utility grid. ESs within the MG can participate in demand response programs, aiding in the stabilization of the utility grid and possibly generating revenue for the MG operator.

4. *Voltage and frequency regulation:* The ESs can respond swiftly to voltage and frequency fluctuations in the MG, ensuring that sensitive equipment and appliances within the MG receive high-quality power.

5. *Utility grid-independent operation:* The ESs allow MGs to operate independently from the utility grid for extended periods of time, which can be

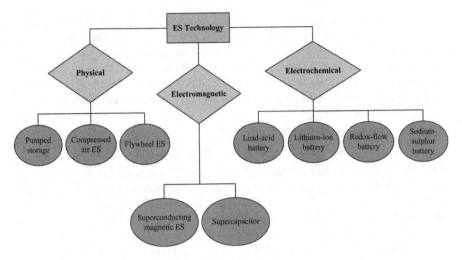

Figure 3.7 Classification of the ES technologies

crucial in remote locations or during emergencies.

3.4 VARIOUS ENERGY STORAGE TECHNOLOGIES

The ES technologies are available in many forms that include physical, electromagnetic, and electrochemical [37, 54]. Further classification of each of these technologies is shown in Fig. 3.6. Following subsections discuss the fundamentals of each of these ES technology.

3.4.1 PHYSICAL ENERGY STORAGE

A physical ES technology converts and stores electrical energy into internal energy or kinetic energy and then converts back the stored energy again into electrical energy as and when required. This technology is indispensable for balancing the intermittent nature of the RESs, controlling peak demand, and providing reserve power. Key physical ES technologies include: (i) pumped storage, (ii) compressed air, and (iii) flywheel. These are discussed in detail below.

3.4.1.1 Pumped storage

A pumped storage technology employs two reservoirs at the storage station, one is located upstream and the other downstream. During the off-peak hours, the ES unit behaves like a motor and pumps up the water stored in the downstream reservoir to the reservoir located at the upstream. When the load demand increases, the same ES unit generates electrical power by allowing the water to flow downstream through a turbine that was previously stored upstream. A pumped storage is capable of supplying electrical power for a time interval ranging from a few hours to a few days without interruption. Operating efficiency of this ES lies between 70% to 85% and it is the most widely used form of ES among all available. By utilizing this ES technology, overall efficiency of the thermal and nuclear power plants can be increased. It is mainly employed for load shifting, emergency reserve, and black-start.

3.4.1.2 Compressed air ES

A compressed air ES (CAES) technology utilizes surplus energy to compress air which is further utilized for electrical power generation by means of a gas turbine. During the off-peak hours, the excess power of the grid is used to compress air which is then stored in an underground reservoir at a pressure of around 7.5 MPa. During the peak hours, when the load demand increases, the compressed air is released to drive a gas turbine to generate electrical power. The response time of a CAES unit is in the range of 1 – 10 mins and it is mainly utilized for load shifting, load balancing, frequency control, power regulation, and emergency reserve.

3.4.1.3 Flywheel ES

A flywheel ES (FES) unit consists of a high-speed flywheel, support bearings, and a motor/generator set. During the off-peak hours, the flywheel is accelerated to higher speed utilizing the excess power from the grid and storing it in the form of kinetic energy. When the load demand increases, the stored kinetic energy of the flywheel is used for generating electrical power through a generator and hence its speed drops. The FES technology is mainly utilized for emergency power reserve, load shifting, and frequency regulation and has a response time of less than 1 s.

3.4.2 ELECTROMAGNETIC ENERGY STORAGE

An electromagnetic ES technology stores energy in the form of electric or magnetic field and discharges to supply electrical energy. Key electromagnetic ES technologies include: (i) superconducting magnetic ES and (ii) supercapacitor. These are discussed below in detail.

3.4.2.1 Superconducting magnetic ES

A superconducting magnetic ES (SMES) technology stores energy in a magnetic field created by flow of direct current in a superconducting coil. The stored energy can be released, when the load demand increases, by discharging the coil. An SMES unit has a quick response time (less than 5 ms), high conversion efficiency, and capability to exchange large amount of energy with the power system in real time. It is principally utilized for voltage control and frequency regulation, thereby, enhancing the stability and transmission capacity of the power system.

3.4.2.2 Supercapacitor

A supercapacitor is a high-capacity capacitor as compared to the conventional capacitor. In general, the storage capacity of a supercapacitor is about 20 to 1000 times that of a conventional capacitor. It is developed based on the electromagnetic double-layer theory where a two-layer capacitor is formed when the electrode is in contact with the electrolyte. However, since the voltage output of a single cell is low, hence to increase the charge/discharge times, several capacitors are connected in either series or parallel in a supercapacitor. The supercapacitor is mainly utilized for short-time and high-power load shifting, voltage and load change compensation, and supply high power instantaneously. It has high power density and high energy conversion efficiency.

3.4.3   ELECTROCHEMICAL ENERGY STORAGE

An electrochemical ES converts chemical energy stored in its active materials into electrical energy. This technology is less prevalent than other energy storage technologies, but it has its own applications and benefits. Principal electrochemical ES technologies include: (i) lead-acid battery, (ii) lithium-ion battery, (iii) redox-flow battery, and (iv) sodium-sulphur battery. These are discussed below.

3.4.3.1   Lead-acid battery

A lead-acid battery is now a mature technology and is widely used for an uninterruptible power supply in the power system, electric vehicles, and communication systems. It normally consists of a lead cathode, a lead dioxide anode, and a low-concentrated solution of sulphuric acid as an electrolyte. Fundamental advantages related to a lead-acid battery are its low cost, high efficiency, and improved reliability. However, its lifetime is short and its manufacturing leads to some environmental hazards since it uses lead which is a poisonous metal. In power system, a lead-acid battery is mainly employed for closing circuit breaker contacts, relay operation, and emergency lighting in case of power failure in substations. Response time of a lead-acid battery is less than 10 s.

3.4.3.2   Lithium-ion battery

Compared to the lead-acid battery, a lithium-ion battery has longer life, low self-discharge rate, higher power density. However, its manufacturing and maintenance costs are very high. But nevertheless, the future potential of this ES technology is expected to be very high as compared to the other battery

ESs especially in the MGs. Currently, the lithium-ion battery is widely used in electronic devices such as smartphones and laptops. Last but not the least, this ES technology is also employed by the Tesla Inc., USA in manufacturing of electric vehicles on a large scale.

3.4.3.3 Redox-flow battery

The redox-flow battery technology offers flexibility is increasing its capacity by either increasing the amount or concentration of its electrolyte. Benefits associated with a redox-flow battery are its long lifetime (about 40 years) and 100% discharge capability. This technology can be counted upon for utilization in an MG for increasing its reliability, stability, load shifting, and emergency power supply.

3.4.3.4 Sodium-sulphur battery

The sodium-sulfur battery technology is still a developing technology. Key advantages of this technology are its high power density, long lifetime (around 10 years), and high efficiency ranging from 80% to 90%. However, if liquid sodium, which is used in this battery, comes in contact with water could prove to be pernicious. Nevertheless, this ES technology can also be considered as a promising means to enhance the security and reliability of an MG, load shifting, and for supplying power during emergency conditions.

A comparative summary of all the ES technologies discussed so far is presented in Table 3.2.

Table 3.2
Comparative summary of the various ES technologies

Type	ES technology	Efficiency (%)	Response time	Advantage(s)	Disadvantage(s)
Physical	Pumped storage	70% - 85%	–	Provide energy balancing, stability, and ancillary services	High initial capital cost
	CAES	40% - 50%	1 - 10 min	large capacity, long continuous discharge time	limited location, slow response
	FES	85% - 95%	< 1 s	High power density, quick response	Low energy density
Electromagnetic	SMES	90% - 95%	< 5 ms	High power density, quick response	Low energy density, costly
	Supercapacitor	95% - 99%	< 1 s	high power density, quick response	low energy density, costly
Electrochemical	Lead-acid battery	80% - 90%	< 10 s	High energy density, large capacity	Low power density
	Lithium-ion battery	≃ 99%	< 10 s	high energy density	small capacity
	Redox-flow battery	65% - 75%	–	long lifetime, low cost	requires expensive fluids that are toxic
	Sodium-sulphur	≃ 90%	< 10 s	high energy density	needs high working temperature

3.5 EXISTING POLICIES TO PROMOTE THE ES TECHNOLOGY DEVELOPMENT IN INDIA

Following are the existing key policies adopted by the Ministry of Power, Government of India to promote the ES technology development in India [55]:

3.5.1 LEGAL STATUS TO THE ES TECHNOLOGY

The Electricity (Amendment) Rules, 2022 say that ES systems are part of the power system, which is what clause (50) of section 2 of the Act means. Also, these Rules say that ES systems can be used by itself or with infrastructure for generation, transmission, and distribution. It will be given a status based on its use area, which is production, transmission, or distribution. ES technology owners or developers can rent or sell storage space to utility companies or Load Despatch Centers (LDCs). They can also use the storage space to buy electricity and keep it for later sale.

3.5.2 ENERGY STORAGE OBLIGATION

The Ministry of Power has announced a long-term plan for Energy Storage Obligations (ESO) on July 22, 2022, to make sure that obligated organizations have enough storage space. The trajectory says that ES technology must provide a minimum percentage of the power used in the area of a Distribution licensee that comes from renewable energy. According to the plan set out in this Ministry's order from July 22, 2022, the ESO of required companies will rise slowly, from 1% in FY 2023–24 to 4% by FY 2029–30, with a 0.5% rise each year. The Appropriate Commission may raise the ESO even more if the planned path, available resources, and needs for grid safety allow it.

3.5.3 REPLACEMENT OF THE CONVENTIONAL DEG UNITS WITH THE RE/ES TECHNOLOGY

The Electricity (Rights of Consumers) Amendment Rules, 2022, which were made public on April 20, 2022, say that people who use the conventional DEG units as backup power must try to switch to cleaner technology like RE with battery storage and the like within five years, or by the date set by the State Commission, depending on how reliable the supply is in the city that is part of the district's supply area.

3.5.4 PROCUREMENT AND UTILIZATION OF THE BATTERY ES TECHNOLOGY

The Ministry of Power, in a resolution from March 10, 2022, laid out specific rules for how to buy and use battery storage as part of assets for generation, transmission, or distribution, or with other services. These guidelines, among other things, make it easier for everyone to buy battery storage in the same way. They also set up a way for different parties interested in buying energy storage and storage capacity to share the risk, which boosts competition and makes these projects more likely to be financed.

3.5.5   RE MUST RUN RULES

On October 22, 2021, the Electricity (Promotion of Generation of Electricity from Must-Run Power Plant) Rules, 2021 were made public. According to these rules, any power plant based on wind, solar, wind-solar hybrid, or hydro source (if there is excessive water leading to spillage) or any other source, as specified by the appropriate government, that has agreed to sell electricity to someone else must be run. It is against the law for a must-run power plant to be limited in its ability to generate or supply electricity because of merit order dispatch or any other business reason, unless there is a technical problem with the utility grid or the grid should be kept safer. In the event that a must-run power plant cuts off its supply, the procurer must pay the must-run power plant compensation at the rates set out in the agreement for the purchase or sale of electricity. These Rules will make it more important for the system to have ES technology so that the RE does not get cut off and penalties are not given out as planned.

3.5.6   ANCILLARY SERVICES FROM ES TECHNOLOGY UNDER THE CERC (ANCILLARY SERVICES) REGULATIONS, 2022

The Central Electricity Regulatory Commission (Ancillary Services) Regulations, 2022 were made public on January 31, 2022. They set up ways for people to buy, use, and be paid for ancillary services at the regional and national levels. The regulations' goal is to keep utility grid frequency within the allowed band and ease transmission congestion so that the grid can work reliably and steadily. Under certain situations, the Regulations say that ES technology can offer Secondary Reserve Ancillary Service (SRAS) and Tertiary Reserve Ancillary Service (TRAS). This will give the ES technology companies another way to make money and encourage investments in energy storage.

3.5.7   BIDDING GUIDELINES FOR ROUND THE CLOCK RE SUPPLY

In November 2020, rules were made public for a tariff-based competitive bidding process to buy round-the-clock (RTC) power from the grid-connected RE power projects, along with power from other sources or storage. These guidelines say that firm power from storage can be used to balance green energy and give 24 hours a day power to buyers/distribution companies (DISCOMs). This helps the State LDCs (SLDCs) keep the utility grid safe and stable in the areas they control. Getting the RTC power will make people want to set up ES technology in the country, which will speed up the energy shift.

3.6 SUMMARY

This chapter provides a comprehensive overview of the various DG sources that are presently available, including their capacity and the accompanying advantages they offer. Subsequently, the significance of the ES in an MG is underscored, followed by a discussion on the several ES technologies now accessible and their multifaceted applications. Furthermore, in order to enhance the understanding of the readers, this study incorporates a comparative analysis and synthesis of diverse distributed generation sources and energy storage systems.

QUESTIONS

1. Define distributed generation (DG). Also, discuss the benefits associated with it from economical and operational viewpoints.
2. Give a detailed classification of the various DG sources utilized along with their key advantages and limitations.
3. Why is the energy storage (ES) technology of significance in a microgrid? Discuss the various ES technologies available along with their principal features.
4. Discuss the existing policies in India to promote the ES technology development in detail.

4 Electric Vehicle

The preceding decade has witnessed unprecedented growth in the number of electric vehicles the EVs running on the road. The primary reason for this is the several inherent advantages related to the EVs be it economical, technical, environmental, or customer convenience. An EV is installed with an electric motor and a battery pack, to power the motor, for its operation. Experts predict that in this coming decade, the majority of the vehicles running on the road will be EVs. In this respect, the following chapter presents a detailed discussion on the classification of EVs, its probable role in the utility grid, and various relevant technologies. Finally, the chapter concludes with a comprehensive discussion on the current status of the EVs and their future trend globally as well as in India.

4.1 OVERVIEW

For about greater than a century, the internal combustion (IC) engine-based vehicles were in dominance in all sorts of transportation sectors. These vehicles used the conventional fuels like gasoline or diesel for their operation. It is noteworthy here that not only these fuels come at a higher price but also emit harmful pollutants (like CO_2) that degrade the environmental health. As a consequence, an accelerated deployment of the EVs has been witnessed in the last decade. Although the EVs came into existence in the midst of the nineteenth century (mainly for public transportation like electric locomotives), the development of electric cars for personal use has stimulated the craze for the EVs over the preceding decade. Figure 4.1 clearly justifies this fact that shows a tremendous rise in the number of EVs worldwide from 2010 to 2020 [56]. Compared to the number in 2010 which was just around 0.03 million, this number has reached approximately 28 million in the year 2022. The EVs shown include light-duty vehicles, light-commercial vehicles, buses, and trucks.

All the EVs comprise an electric motor with an efficiency ranging in between 90% to 95%. Various types of motors that are commonly utilized are brushless DC motors, switched reluctance motors, and permanent magnet synchronous motors [57]. To power the motor and make it run, a battery bank is also installed within the EV. Usually lithium-ion batteries are employed in the present-day EVs because of their high energy density, high specific energy, light weight, long life, compactness, increased durability, and fast charging. However, researchers are working on developing lithium-air, lithium-oxygen, and ultracapacitors to replace the lithium-ion batteries in the future owing to their inherent advantages. The Key benefits associated with the EVs are zero tailpipe emissions resulting in reduced air pollution and noise-free oper-

DOI: 10.1201/9781003477136-4

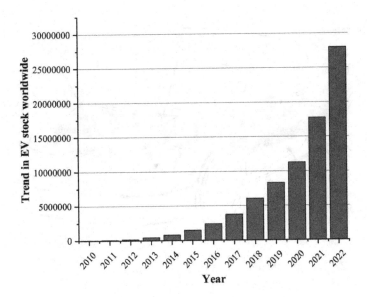

Figure 4.1 Trend in the EV stock worldwide from 2010 to 2022 (Source: IEA Tech. Report 2023)

ation, compared to the conventional IC engine-based vehicles, reducing noise pollution.

MGs are believed to be the future of power systems. Sooner or later the conventional power systems would be replaced by the MGs that will generate power locally and consume it locally as well. WTGs and solar PV modules will hold a major share for generating power in the MGs. But since the power output from these is unpredictable and intermittent, owing to their dependence on weather conditions, EVs can play a significant role in their integration in the MGs and increase the reliability of the power output from these by: (i) using battery banks of the EVs as mobile energy storage system, (ii) utilizing second hand EV batteries as stationary energy storage system, iii) deployment of charging technologies and infrastructure on a large scale, (iv) display of a supportive attitude by the EV owners, and (v) providing ancillary services to the utility grid. The ancillary services can be provided by using the battery of the EVs to store excess power and release it when required to assist in frequency regulation, shaving peak load demand, and power support to the MG to maintain its reliability and stability [58].

Figure 4.2 shows the layout of major components installed within an EV. These include: (i) a battery bank, (ii) an electric motor, (iii) a converter, (iv) charging port, and (v) transmission system. As outlined above, the battery bank in modern EVs consists of lithium-ion batteries, although research is going on to replace these by the lithium-air, lithium-oxygen, and

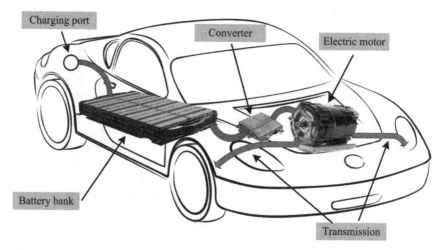

Figure 4.2 Layout of major components of an EV

ultracapacitor based batteries. An electric motor may be either ac (permanent magnet synchronous motor) or dc (brushless dc motor or switched reluctance motor). A converter, depending upon the type of motor utilized, may be either dc to dc converter or ac to dc converter. The charging port consists of a small ac to dc converter that is used to charge the battery bank placed in the EV. The power output of the electric motor in an EV is transferred to its front wheels to rotate them by the transmission system.

4.2 CLASSIFICATION OF THE EVs

The EVs can be classified into two main types. These are battery EVs (BEVs) and plug-in hybrid EVs (PHEVs) [58]. Each of these is discussed below in brief.

4.2.1 BATTERY EVs

The BEVs totally rely on rechargeable batteries for their operation and no other means (like IC engines, fuel cells, etc.) are utilized. For an uninterruptible service, the batteries must be recharged once a BEV finishes one trip. Compared to the PHEVs, the BEVs employ a battery bank of larger capacity. The battery bank of a BEV is directly charged using electricity from the grid. The charger efficiency ranges between 60% and 90%. Combined with the electric motor efficiency (90% to 95%), the overall efficiency of a BEV is achieved around 70% to 75% [59]. Compared to a conventional IC engine-based vehicle with an efficiency of 20% to 30%, the BEV possesses a much greater efficiency alongside providing a comparable design, comfort, and safety. Most popular modern-day BEVs include Nissan Leaf, Renault Zoe, Chevrolet Bolt, Tesla

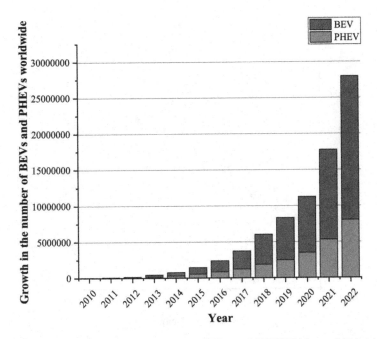

Figure 4.3 Growth in the number of BEVs and PHEVs from 2010 to 2022 (Source: IEA Tech. Report 2023)

Model S, Hyundai Kona Electric, Porsche Taycan, Tesla Model Y, and Tesla Model 3.

4.2.2 PLUG-IN HYBRID EVs

The PHEVs encompass both a battery bank and an IC engine. The IC engine is ignited from a liquid fuel stored in a tank within the EV. Unlike the BEVs, the PHEVs do not have a charging port for recharging the battery bank instead the bank relies on the liquid fuel for its recharging. The battery bank used is of smaller capacity compared to that used in the BEVs since liquid fuel is also used. The combination provides the PHEVs with a longer driving range. As per the technical report on EVs released by IRENA in 2017 [58], the PHEVs can achieve a driving range of 750 km or more (with battery bank as well as liquid fuel) as compared to the driving range of the BEVs that lies between 300 km and 400 km. Some of the most-selling PHEVs worldwide include Honda Clarity, Toyota Prius Prime, Ford Escape, Mitsubishi Outlander, Hyundai Santa Fe, and Kia Niro.

Figure 4.3 shows the growth in the number of BEVs and PHEVs from 2010 to 2022 [56]. Compared to only 27400 BEVs and 610 PHEVs in the year 2010, the year 2022 shows a massive growth in their numbers with 19.93 million BEVs and 8.04 million PHEVs. Figure 4.4 shows a comparative

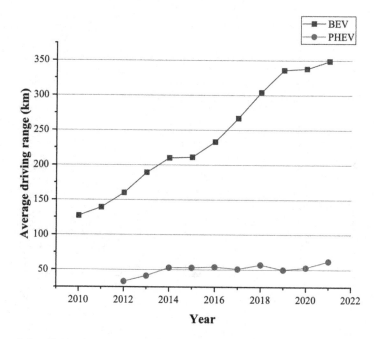

Figure 4.4 Comparative average driving range of BEVs and PHEVs over the years 2015 – 2020 (Source: IEA Tech. Report 2021)

average driving range of the BEVs and the PHEVs (considering only battery bank-dependent driving) by powertrain from 2010 to 2021 [60]. It can be said that the BEVs have a much greater potential of driving longer distances as compared to the PHEVs with the average driving range of the BEVs around 350 km and the average driving range of the PHEVs below 100 km for the year 2021.

4.3 MODES OF INTERACTION OF THE EVs WITH THE UTILITY GRID

Now it is well understood that integration of the EVs with the grid can facilitate a greater penetration of the unpredictable and intermittent RES-based generators in the grid. Another positive aspect of the EV integration with the grid is that it will result in decarbonization of the power system and thus will result in an improved environmental health. The question here arises is that what are the various means by which the EVs can be integrated with the grid or how can they interact with it. Some of the important modes of interaction of the EVs with the utility grid are vehicle-1-grid (V1G), vehicle-to-grid (V2G), vehicle-to-customer (V2C), and vehicle-to-X (V2X) [61]. Although all of them except one are still in testing phase or have been implemented only in countable commercial applications [62].

4.3.1 V1G MODE

The V1G mode refers to the case when the EV battery pack is being charged (unidirectional charging) only by consuming the grid power. The battery pack needs to be charged after an EV finishes a trip and before getting ready for the next trip. In general, it is advisable to charge the EV battery pack at late nights when the power demand is not high. This will help the utility grid in not generating extra power during the peak-load hours, thus maintaining the reliability and stability of the grid operation. The V1G mode is a part of the demand side management whereby the EV user can choose when to charge the EV battery. To promote this, the utility offers certain financial incentives, cheaper energy prices, and other benefits to the consumers for charging their EVs during the off-peak hours.

4.3.2 V2G MODE

The V2G mode refers to the case where the EV battery pack observes bidirectional flow of power with the utility grid. This is to say that unlike the V1G mode where the EV battery is only in charging mode, the V2G mode allows the EV battery to get charged by utilizing the grid power as well as discharge by supplying power to the grid when required. This allows for an enhanced penetration of the RES-based power generators in the grid especially in the MGs. Although the V2G mode of interaction is still immature and is in testing phase, it has every possibility of emerging as a potential solution in ensuring a stable and reliable operation of the utility grid. The V2G mode will allow the consumers to charge their EV battery whenever required or else help the grid by supplying the power stored in the battery during the peak-load hours.

4.3.3 V2C MODE

The V2C mode includes the vehicle-to-building (V2B) mode and the vehicle-to-home (V2H) mode. The V2B mode refers to the bidirectional operation of the EVs integrated with the non-residential buildings providing behind-the-meter services whereas the V2H mode refers to the bidirectional operation of the EVs integrated with the residential buildings providing behind-the-meter services. Both these modes are currently in an experimental phase.

4.3.4 V2X MODE

The V2X mode may be either V2G mode or V2C mode. Researchers in [61] have discussed various technical, social, and regulatory aspects of this mode of interaction with the grid. The technical aspects examine the factors affecting the battery health degradation that includes battery chemistry and size,

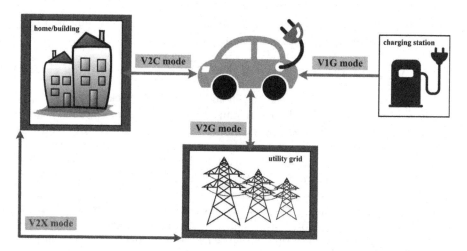

Figure 4.5 Various modes of interaction of the EV with the utility grid

temperature, driving, and battery charging behavior. The social aspects investigate the charging pattern of the EV users during the whole day. Currently
the investigation related to the social aspects is not mature enough and thus
needs to be prioritized. The regulatory aspects explore the barriers related to
the market participation of the small capacity service providers and lack of
definitions for the ES technologies. A detailed analysis of all these aspects can
be located in [61].

A pictorial representation of all these interaction modes is shown in Figure
4.5.

4.4 IMPLEMENTATION OF THE EVs FOR THE LFC

As already discussed in Chapter 1, mismatch between power generation and
load demand in a power system prominently affects its reliability and stability.
A permanent disparity between these two may result in undesirable and large
deviations in the system frequency that may ultimately damage the costly
equipment, degrade load performance, overload transmission lines, and even
lead to a complete power system blackout. The LFC in such a circumstance
plays an indispensable role by holding up the power generation with the load
demand, thereby, maintaining a stable frequency profile. In addition, increasing size of the power system, its changing structure, and integration of various
DG technologies with unpredictable and intermittent operation (especially in
the case of an MG) demand for an effective LFC scheme to be implemented.
Various ES technologies are integrated into the MG to mitigate the effects of
the RES-dependent power generators on the system frequency. It is noteworthy here that compared with the conventional battery energy storage system,
the battery pack of an EV possesses lower degradation tendency and lesser

cost [20, 63]. This fact makes the EVs to be a potential candidate to be integrated into the MG for the frequency control process as an alternative to the conventional battery energy storage technology. Countless works can be spotted in the quality literature over the past decade that justify a successful implementation of the EVs for the LFC analysis of both the conventional power system and various configurations of an MG [10–14, 17, 23, 63–70].

The EV is considered to participate in the LFC process depending upon the state-of-charge (SoC) of its battery. The SoC of a battery is defined as [71]:

> *The capacity that is currently available as a function of its rated capacity*

A 100% SoC means that the battery is fully charged, whereas a 0% SoC means that the battery is completely discharged. Generally, the battery SoC is expressed in percentage. A lower level of the battery SoC (SoC_{min}) and an upper level of the battery SoC (SoC_{max}) are specified by the central load dispatching center abandoning which the EV is not allowed to contribute in the frequency control process. Reasons being, respecting the EV users' convenience and ensuring a long battery life. Most of the past research works have considered $SoC_{min} = 80\%$ and $SoC_{max} = 90\%$ for the LFC analysis of power systems [10, 14, 17, 63, 69, 72–75]. Although other values have also been considered for these [16, 76–79].

It is to be highlighted here that the EVs participate in the LFC process by utilizing the V2G mode of interaction or the V2G technology. The above cited works consider the V2G technology for the LFC analysis. As already stated, compared with the V1G technology that emphasizes only on unidirectional flow of power from the grid to vehicle, the V2G technology embraces a bidirectional power flow, i.e., from the grid to vehicle (charging of the EV battery) and from the vehicle to grid (discharging of the EV battery). When the load demand increases, the EV shifts to discharging mode thus acting as a power source and when the load demand falls, the EV shifts to the charging mode thus acting as a load to the grid [67]. A graphical comparison between the V1G and the V2G modes of interaction of the EV with the grid is shown in Fig. 4.6.

While implementing the V2G mode, the EVs interact with the grid through two different SoC control schemes [10]. These are the *synchronous SoC control scheme* and the *asynchronous SoC control scheme*. Both these schemes differ from each other in a way the LFC signal is dispatched to the EVs for charging/discharging of their batteries. A number of local control centers are assigned to monitor the SoC of a group of the EV batteries. All these local control centers are further controlled by a central load dispatching center that decides the dispatching of the LFC signal to the EVs. The LFC signal dispatching process is shown in Fig. 4.7. The schemes are briefed below.

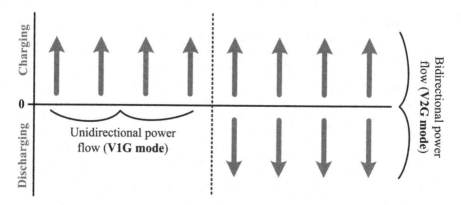

Figure 4.6   The V1G vs the V2G modes of interaction of the EV with the utility grid

4.4.1   SYNCHRONOUS SOC CONTROL SCHEME

In this scheme, the central load dispatching center dispatches the LFC signal to the local control centers. The local control centers further distribute the incoming LFC signal to the group of the EVs they are monitoring. The SoC of the EVs being controlled by a particular local control center becomes synchronized and the EV batteries are taken as a virtual battery energy storage system by the central load dispatching center. The LFC charging signal

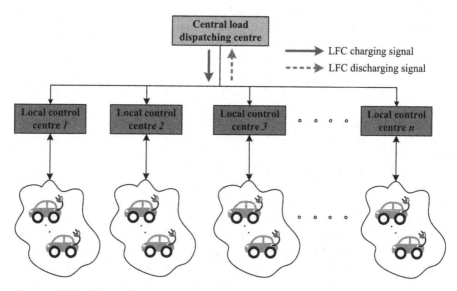

Figure 4.7   Dispatching of the LFC signal for charging/discharging of the EV batteries

is dispatched to the EVs in ascending order of their battery SoC whereas the discharging LFC signal is dispatched based on the descending order of the battery SoC. The charging/discharging priorities are decided in every 30 seconds.

4.4.2 ASYNCHRONOUS SOC CONTROL SCHEME

In this scheme, the SoC of the EV batteries monitored by a particular local control center is not synchronized. No charging/discharging priorities are considered in this scheme. Whatever the LFC charging/discharging signal is dispatched from the central load dispatching center, it is equally distributed among all the local control centers.

4.5 CURRENT STATUS AND FUTURE TRENDS OF THE EVs

This section discusses in detail the current status and future trends in the implementation of the EV technology. Both Indian and global perspectives are undertaken and examined. This section makes sense in a way that after a detailed discussion on the EV technology, their types, interaction modes with the utility grid, and their role in the LFC analysis of power system, it becomes imperative to discuss the current status of the E technology and its future trends.

4.5.1 GLOBAL PERSPECTIVE

As per the Stated Policies Scenario, IEA report of 2023 [56], the global stock of the EVs (considering both private and public) at the end of the year 2022 was around 27 million (excluding two and three wheelers). The leading nations in this list were China, France, Germany, Japan, Spain, United States of America (USA), and the United Kingdom (UK). Figure 4.8 shows the stock of EV cars (both BEVs and PHEVs included) at the end of the year 2022 [56]. It can be seen that China leads the list with around 4.5 million BEVs and around 1 million PHEVs followed by the USA. Table 4.1 shows the stock of EV buses, vans, and trucks (both BEVs and PHEVs included) in the same selected countries in the year 2022. Here also China leads the group in every segment. The United States has implemented none of these transport modes yet. However, France, Germany, and the UK are showing considerable progress in these.

Figures 4.9 to 4.12 show a comparative future trend in the stock of cars, buses, vans, and trucks, respectively, till the year 2030 in China, Europe, USA, and rest of the world. Both the BEVs and the PHEVs are included in the stock. As can be observed China will continue to lead in the stock of cars, buses, and trucks except for the vans. These trends are as per the Stated Policies Scenario, IEA report of 2023 [56].

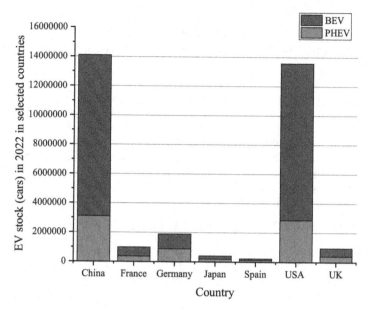

Figure 4.8 The EV car stock (BEVs and PHEVs included) at the end of the year 2022 in selected countries around the world (Source: IEA Tech. Report 2023)

4.5.2 INDIAN PERSPECTIVE

As per the Stated Policies Scenario, IEA report of 2023 [56], the total BEVs and PHEVs (including cars, buses, vans, and trucks) in India at the end of the year 2020 were 16800 and 50, respectively. This is shown in Fig. 4.13. The same figure also shows the future trend of these till the year 2030. At the end

Table 4.1

Stock of EVs (buses, vans, and trucks) in selected countries in 2022

Country	Buses		Vans		Trucks	
	PHEV	BEV	PHEV	BEV	PHEV	BEV
China	100000	670000	1700	460000	21000	290000
France	70	1300	990	74000	0	81
Germany	130	1700	770	53000	0	900
Japan	0	77	0	39000	0	55
Spain	100	610	200	14000	0	200
USA	0	0	0	0	0	0
UK	220	1800	2700	39000	0	950

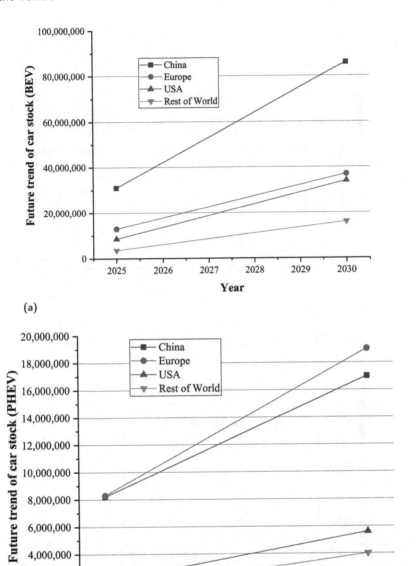

Figure 4.9 Comparative future trend of cars (BEVs and PHEVs included) till the year 2030 in China, Europe, USA, and the rest of the world

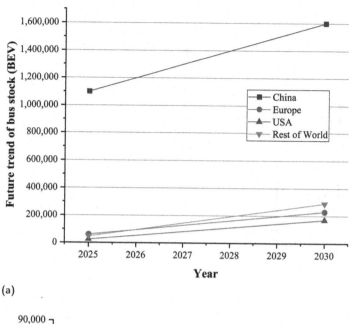

(a)

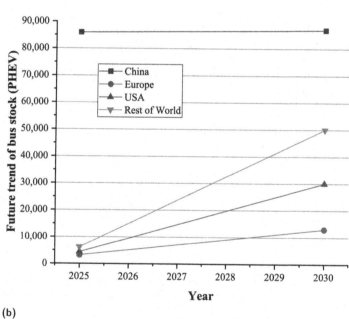

(b)

Figure 4.10 Comparative future trend of buses (BEVs and PHEVs included) till the year 2030 in China, Europe, USA, and the rest of the world (Source: IEA Tech. Report 2023)

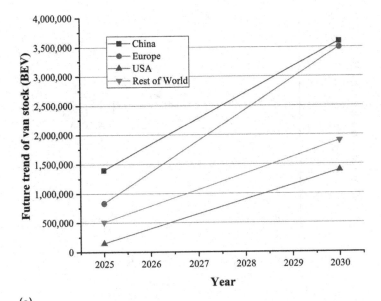

(a)

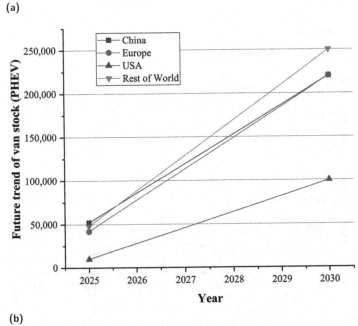

(b)

Figure 4.11 Comparative future trend of vans (BEVs and PHEVs included) till the year 2030 in China, Europe, USA, and the rest of the world (Source: IEA Tech. Report 2023)

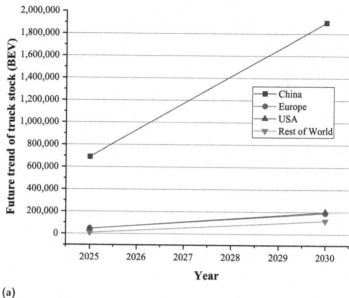

(a)

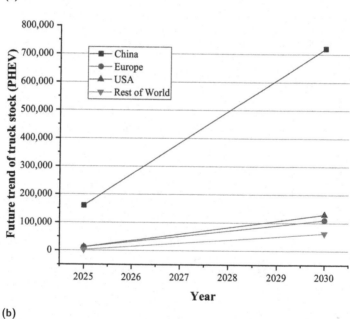

(b)

Figure 4.12 Comparative future trend of trucks (BEVs and PHEVs included) till the year 2030 in China, Europe, USA, and the rest of the world (Source: IEA Tech. Report 2023)

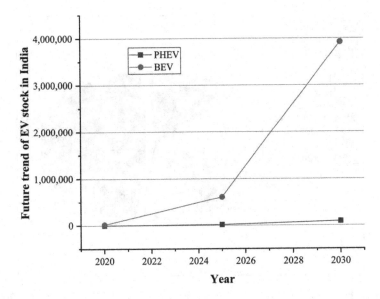

Figure 4.13 Status of the EV stock in the year 2020 and future trend till the year 2030 (BEVs and PHEVs included) in India (Source: IEA Tech. Report 2023)

of 2030, the BEVs and the PHEVs are expected to be around 3919000 and 91410, respectively. Figure 4.14 presents the monthly EV sale (four wheelers) in India (excluding the sale in Telangana state) between October 2022 and September 2023 [80]. The EV sales account for a total of 43740. A breakage in the future trend of cars and buses, vans, and trucks is shown in Figs. 4.15 (a)-(b), respectively. It can be stated that compared to the buses, vans, and trucks, the cars are expected to rise at a much greater pace.

4.5.3 IMPORTANT POLICIES TO PROMOTE THE DEPLOYMENT OF THE EVs IN THE FUTURE

Governments across the globe have announced and implemented certain policies to encourage the deployment of the EVs in the future. Some of the important policies include [56]:

1. *The New Energy Automobile Industry Plan (NEAIP):* The NEAIP, implemented in China for the period 2021 to 2035, targets 20% of the total vehicle sales to be zero emission vehicles (ZEVs) by 2025. Furthermore, EV sales over 50% are targeted by the year 2035.

2. *EU Green Deal:* The EU Green Deal targets to push the average tailpipe CO_2 emissions/km below 95 gm in 2021. Further, the emissions are targeted

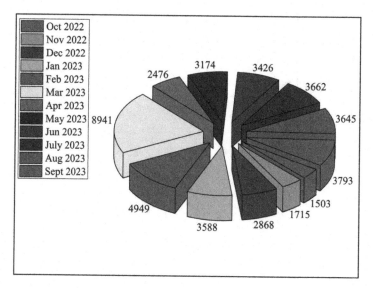

Figure 4.14 Sale of the EVs (four wheelers) in India from October 2022 to September 2023

to be reduced by 15% in 2025, by 37.5% in 2030 and adoption of well-to-wheel approach rather than the current tailpipe (tank-to-wheel) approach.

3. *Faster Adoption and Manufacturing of Electric Vehicles (FAME II):* The FAME II scheme is India's key policy that allocated USD 1.4 million for three years i.e. 2019 to 2021 for a stock of 1.6 million PHEVs and BEVs (including two/three wheelers, buses, and cars). The scheme also aimed at manufacturing of EVs and their parts within the country. Further, the scheme targets a sale of 30% EVs among the total on-road vehicles by the year 2030.

4. *Green Growth Strategy:* In Japan, the Ministry of Economy, Trade and Industry (METI) proposed a Green Growth Strategy that aims at carbon neutralizing the whole of Japan by 2050. By the mid-2030s, the strategy targets to have all the new passenger cars electrified. Also, Japan doubled its subsidies in 2020 on the sale of EVs across all transport modes.

5. *National Electromobility Strategy:* Chile's National Electromobility Strategy targets to achieve a ten-fold increase in the number of EVs in 2022 compared to 2017. The strategy also aims to increase the penetration rate of the private EVs to 40% by 2050 and that of the public EVs to 100% by 2040.

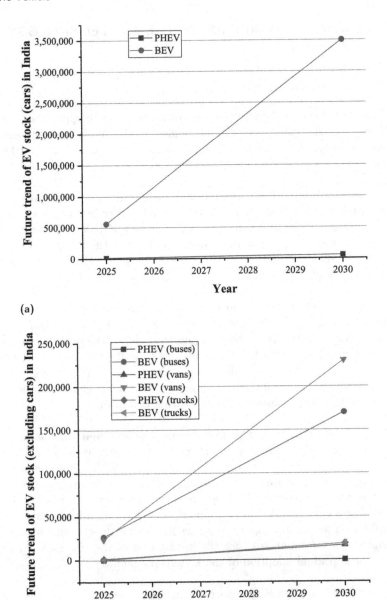

(a)

(b)

Figure 4.15 Breakage of the future trend in a) the EV car stock and b) the EV bus, van, and truck stock till the year 2030 in India (Source: IEA Tech. Report 2023)

4.6 DOMINANT BARRIERS IN THE EV TECHNOLOGY DEPLOYMENT

A successful practical deployment of the EV technology is supposed to face the following barriers [81–83]:

1. *High capital cost:* Compared to the conventional IC engine-based vehicle, the capital cost of an EV is much higher. Furthermore, the maintenance cost and battery cost of the EV incur additional costs. This makes the low- and mid-income class reluctant to afford an EV.

2. *Low driving range:* An EV has a lower driving range compared to a conventional IC engine-based vehicle. As discussed earlier in section 4.2, a BEV has a driving range of 300 to 400 km, whereas a PHEV possesses a higher driving range of around 750 km. But compared to the IC engine-based vehicle, this driving range is still low. As a result, an EV user needs to be selective about trips and may not be able to plan long trips.

3. *Battery charging time:* In general, slow and fast chargers are available for charging an EV battery. Slow chargers usually take 7 to 8 hours to fully charge the battery. Also, the charging time depends upon the size of the EV battery. Larger the battery size more time it takes to charge. On the other hand, a fast charger takes comparatively less time to charge the battery. However, charging rate of the battery with a fast charger reduces with decrease in temperature or in cold weather conditions.

4. *Charging infrastructure:* This factor plays a crucial and indispensable role in adoption and deployment of the EV technology. Since, an EV battery needs to be charged after every long trip, it is a must requirement for the charging infrastructure to be installed at diverse places to help the EVs in charging their batteries. Although in developed countries like the United States, the UK, Japan, and many more, the charging infrastructure is well established but in developing countries like India, the government needs to formulate strong mandates for the installation of the charging infrastructure for a gradual adoption of the EVs in the country.

5. *Public awareness for EV support:* To increase the adoption of the EVs, it is mandatory to make the public aware of the benefits associated with the EVs like environmental benefits, fuel savings, and the various incentives and schemes that the government offers for them. Providing knowledge to the people about the EVs through educational programs, questionnaires, and interactive sessions could be of great help in curbing the attitude of the people towards the utilization of the EV technology.

6. *Vehicle servicing:* Since the EV technology is different from a conventional IC engine-based technology, it demands a skilled technician for maintenance, servicing, and troubleshooting of the EVs.

It requires a concerted effort from governments, industry stakeholders, and consumers to overcome these obstacles. Infrastructure development, policy support, incentives, and increased public awareness are all crucial for the widespread adoption of the EV technology. Furthermore, substantial progress in charging infrastructure and battery technology will be required to combat these obstacles.

4.7 SUMMARY

The chapter commences with providing an introductory examination of the EV technology, followed by a comparative analysis of conventional IC engine-based vehicles. This comparison highlights the advantages of electric vehicle technology and underscores its remarkable growth over the previous decade. A wide range of EVs, together with their associated advantages and disadvantages, have garnered significant interest in this context. Subsequently, a thorough examination was conducted on the potential integration of the EVs with the utility grid, encompassing a wide range of strategies for facilitating EV-grid interaction. The role of the EVs in the frequency control of the grid has been acutely emphasized. The conclusion of this study incorporates an analysis of both the global and Indian perspectives regarding the current state and future trajectory of the EVs. In summary, this paper has delineated and examined the main barriers that impede the extensive implementation of the EV technology.

QUESTIONS

1. The EVs are considered to be the future of the automotive industry. Elaborate.
2. Discuss the general layout of an EV mentioning the function of each comprising subsystem.
3. Give a detailed classification of the various types of EVs along with relevant examples.
4. Emphasize on the various operating modes of an EV giving a graphical representation.
5. Discuss the significance of the EVs from the LFC point of view of an MG.
6. Discuss the global policies implemented to promote the future deployment of EVs.
7. What are the dominant barriers identified in the EV technology deployment? Discuss.

5 Microgrid

Since the preceding two decades, conventional power plants have been unable to keep up with the steadily growing demand for electricity in the community. As a result, the MGs have arisen as an attractive alternative to the conventional power plants. Reduced greenhouse gas emissions, inclusion of the RESs, and abbreviated transmission losses and feeder capacity are the primary reasons that have accelerated the emergence of the MGs. The MGs are currently being investigated as a potential route to improve the health of the environment as well as the economy and energy requirements. There are several effective applications of an MG that can be found all over the world, including in the USA, Japan, the European Union, and China. In this chapter, the fundamental idea behind an MG, as well as certain standard definitions of it, as well as its composition, modes of operation, and classification, are discussed.

5.1 CONCEPT OF A MICROGRID

An MG is composed of two terms: *micro* + *grid*. *Micro* means very small and *grid* means an entity that aggregates generation, transmission, distribution, and consumption of electrical power. It implies that an MG is a very small power system (of capacity up to a few MW) that facilitates generation, transmission, distribution, and consumption of electrical power in a local area. The local area could be a small household, a public facility (like school, hospital, etc.), a commercial entity, or an industrial site. In the year 2001, Professor R. H. Lasseter of the University of Wisconsin–Madison came up with the fundamental idea that would eventually become the MG. In later years, multiple standard definitions of the MG were proposed by the US Department of Energy, the European Union (EU) Commission Project Microgrid, and the Consortium for Electric Reliability Technology Solutions (CERTS) (CERTS). Since the beginning of the twenty-first century, the MGs have been witness to an increased emergence in the process of formulating a low-carbon economy and promoting the environmental health in order to attain sustainable development. Fewer emissions of greenhouse gases, incorporation of the RESs, lowered transmission losses, and increased feeder capacity are the key factors that have contributed to their fast proliferation. MGs are installed and operated on the MV/LV distribution side, which enables them to allow a bidirectional flow of electrical power on the distribution side. This is in contrast to a typical power system, which assists a unidirectional flow of the electrical power on the distribution side. As a result of the myriad of benefits that are inherently connected with the MGs, it is not an inaccurate assumption to make at this point that MGs are likely to play a significant role in the

DOI: 10.1201/9781003477136-5

evolution of our current power system.

5.1.1 SOME STANDARD DEFINITIONS

Following are some standard definitions given for an MG by the various renowned bodies around the world.
The CERTS defines an MG as [84]

> *A cluster of the DG sources and loads operating as single system providing both power and heat to a particular area*

The majority of the distributed generation sources must be power electronic based in order to operate flexibly as a single aggregated system.

As defined by the U.S. Department of Energy [85]

> *A microgrid is a group of interconnected loads and DG sources with clearly defined electrical boundaries that act as a single controllable entity with respect to the grid and can connect and disconnect from the grid to enable it to operate in either grid-connected or islanded modes*

As per the EU Commission Project [86]

> *A microgrid comprises LV distribution systems with DG sources together with storage devices and flexible loads. Such systems can be operated in a non-autonomous way, if interconnected to the grid, or in an autonomous way, if disconnected from the main grid. The operation of microsources in the network can provide distinct benefits to the overall system performance, if managed and coordinated efficiently*

As per Galvin Electricity Initiative [87]

> *Microgrids are modern, small-scale versions of the centralized electricity system. They achieve specific local goals, such as reliability, carbon emission reduction, diversification of energy sources, and cost reduction, established by the community being served. Like the bulk power grid, smart microgrids generate, distribute, and regulate the flow of electricity to consumers, but do so locally*

5.2   COMPOSITION

An MG basically comprises of various DG sources, ES units (ESUs), control units, and connected electrical loads. Each of these entity is discussed in brief below.

5.2.1   DG SOURCES

The process of generating power locally at the distribution or consumer side, without involving the conventional utility grid, is termed as DG. Various DG sources commonly employed have already been discussed in Chapter 3 of this book. Numerous benefits related to the DGs, be they from economical or operational viewpoint, have also been outlined in the same chapter.

It is noteworthy here that although, innumerable benefits are related with the DG sources but at the same time owing to their different voltage levels and intermittent power outputs, it sometimes becomes difficult to maintain a synchronism between them. This is pernicious to the system operation and stability.

5.2.2   ENERGY STORAGE UNITS

The ESUs are an integral part of an MG that influences its reliability and stability. Various types of ESUs have already been dealt with in Chapter 3 of this book along with their technical specifications. Key functions of the ESUs are summarized below:

1. The ESUs compensate for the intermittency in the power outputs of the weather-dependent units (such as solar power, wind power, etc.) installed in the MG through their charging/discharging and thus ensure an uninterruptible power supply to the consumers.
2. Due to their fast response time, the ESUs assist the various DG sources to counterbalance any abrupt load deviations in the MG and thus sustain a balance between the power generation and load demand.

5.2.3   CONTROL UNITS

Control units refer to the various PECs that are installed in the MG. The PECs facilitate the MG in transition between the grid-connected mode and the islanded mode and thus, tend to manage the power exchange and stability during the transition process. These modes of operation will be discussed in section 5.3. The role of the PECs during the grid-connected and islanded modes is briefed below:

1. In the grid-connected mode, the PECs connect and operate the various DG sources in parallel with the utility grid and aid in exchange of real and reactive powers between the utility grid and the MG.

2. In the islanded mode, the PECs are bound to maintain the voltage and frequency references stable and ensure an uninterruptible power supply to the critical loads connected within the MG.
3. Additionally, the PECs control the charging and discharging process of the various ESUs in the MG. This helps the ESUs to compensate for any abrupt load variations in the MG, thereby, improving the power quality, reliability, and stability of the MG.

5.2.4 ELECTRICAL LOAD

The electrical loads fed by an MG consist of important or critical loads (such as army base, hospitals, etc.) and unimportant or non-critical loads (such as the conventional domestic load). During outage or maintenance of the utility grid, a steady supply to the important loads is to be ensured.

5.3 MODES OF OPERATION

An MG can be operated either in grid-connected mode or islanded mode. Figure 5.1 shows the scheme for transition between the grid-connected mode and islanded mode. The transition between these two modes is carried out through a static switch installed at the point of common coupling (PCC). Both these operating modes of the MG are discussed in brief below [37].

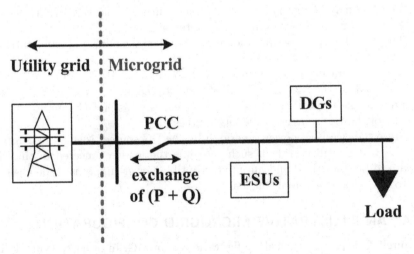

Figure 5.1 Transition scheme of the microgrid between grid-connected mode and islanded mode through PCC

5.3.1 GRID-CONNECTED MODE

When the static switch at the PCC is closed the MG operates in the grid-connected mode. The key points relevant to this mode of operation are summarized below:

1. In this mode, the MG can exchange real (P) and reactive (Q) powers with the utility grid.
2. In case of power deficit, the utility grid derives power from the MG, whereas with surplus power with the utility grid, the loads connected within the MG are fed. This ensures a power balance in the MG.
3. The utility grid controls the voltage and frequency references (V/f control) of the entire system whereas the MG is allowed to only exchange the real and reactive powers (P/Q control) with the utility grid.
4. Output powers of the various DG sources are utilized to the maximum possible extent for an optimum economic operation of the MG.
5. In short, optimization of the various DG sources and coordination between the utility grid and the MG are obligatory in this mode of operation in order to achieve maximum energy efficiency.

5.3.2 ISLANDED MODE

When the static switch is opened the MG starts operating in the islanded mode. The key points relevant to this mode of operation are summarized below:

1. The MG may be disconnected from the utility grid via the PCC in case of a grid failure or as scheduled for maintenance.
2. The islanding may be intentional (preplanned islanding) or unintentional (unplanned and uncontrollable islanding).
3. In this operating mode both the V/f control and P/Q control are to be managed by the MG itself. Out of the total connected DG sources, some provide the voltage and frequency references (master DG sources) to the rest of the DG sources (slave DG sources).
4. Since the electrical power generated by the DG sources is small in comparison to the utility grid, in such a regime the loads connected to the MG are to be prioritized and an uninterruptible power supply is to be ensured to the critical or important loads.

5.4 AN ILLUSTRATIVE MICROGRID CONFIGURATION

Figure 5.2 displays a typical configuration of an MG. In order to synchronize the MG with the utility grid, a voltage level of 11 kV/415 V is employed. To facilitate the seamless transition from grid-connected mode to islanded mode, a PCC has been implemented to establish a connection between the utility grid and the MG. The 415 V MG bus serves as the origin for three radial feeders,

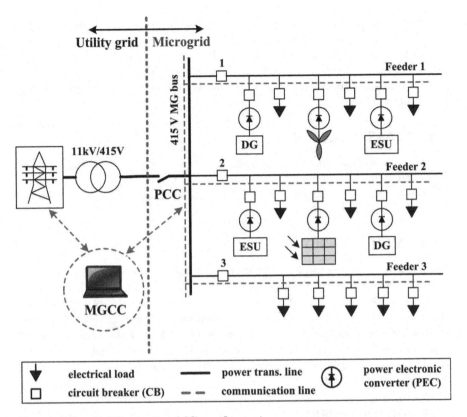

Figure 5.2 An illustrative MG configuration

specifically identified as feeder A, feeder B, and feeder C. Each of these feeders is equipped with a circuit breaker (CB) labeled as 1, 2, and 3, respectively. This enables the feeders to be easily connected or disconnected from the MG bus as required. Feeder 1 and feeder 2 are connected to both the DG sources and the ESUs as well as the RES-dependent generators. The DG sources can be DEG, FC, MT, etc., whereas the ESUs can include BESU, FESU, RFBU, etc., as already discussed in Chapter 3 of this book. Furthermore, the feeders are interconnected with the critical loads to provide an uninterrupted provision of electrical power. Feeder 3 assumes the responsibility of providing power to the linked loads that are deemed non-critical in nature.

The MG central controller, commonly referred to as the MGCC, assumes the responsibility of load management within the MG. The successful control of generation-load balance in the MG is ensured by this approach. The determination of whether the total power generation from all the DGs in the MG is adequate to satisfy the load demand is made by the MGCC. If the generation is deemed sufficient, the MGCC issues instructions to the MG to curtail some non-critical loads. The MGCC effectively governs both the voltage and frequency parameters to uphold the stability of the system. Furthermore, the MGCC assumes the responsibility of monitoring the power quality at the PCC and determining the initiation of the islanding operation for the MG. Following the completion of utility grid repairs, the MGCC will undertake re-synchronization procedures to restore voltage and frequency alignment with the utility grid [88].

5.5   CLASSIFICATION

A very common classification of the MGs is based on supply as AC MG, DC MG, and hybrid MG. Another classification is based on the location of the MGs as: urban MG and remote MG [89]. These classifications are discussed in the following subsections.

5.5.1   AC MICROGRID

In an AC MG, that is most widely used, all the DG sources, ESUs, and loads connect with each other through an AC bus. The DG sources with AC power output (such as the WTG, DEG, MT, etc.) connect to the AC bus via AC/AC PECs for a stable coupling. The DG sources with DC power output and the various ESUs connect to the AC bus via DC/AC PECs. Furthermore, the AC loads connect to the AC bus directly while DC/AC PECs are required for the DC loads in order to connect to the bus. Figure 5.3 shows the general structure of an AC MG.

5.5.2   DC MICROGRID

In a DC MG, each DG source, ESU, and load is interconnected with other nodes in the network over a DC bus. The process of connecting to the DC bus is achieved through the utilization of AC/DC PECs by the DG sources that generate AC power. DC/DC PECs function as the intermediary between the bus and the several DG sources and ESUs that generate DC power. Regarding the loads, AC loads establish a connection with the DC bus via utilizing AC/DC PECs, whereas DC loads directly interface with the bus. It is vital to acknowledge that the connection between the DC bus and the utility grid is established through the utilization of a DC/AC PEC. The diagram presented in Fig. 5.4 illustrates the general configuration of the DC MG, providing an overview of its overall architectural design.

5.5.3   HYBRID MICROGRID

A hybrid MG consists of both the AC and DC buses. The DG sources with DC power output connect to the DC bus, whereas the DG sources with AC power output connect to the AC bus. A general structure of a hybrid MG is shown in Fig. 5.5.

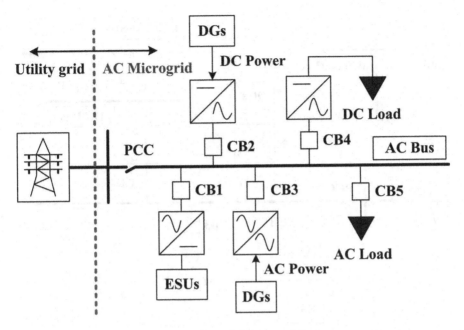

Figure 5.3 General structure of an AC microgrid

5.5.4 URBAN MICROGRID

An urban MG is in general operated in grid-connected mode. As was mentioned previously, when operating in grid-connected mode, an MG is able to

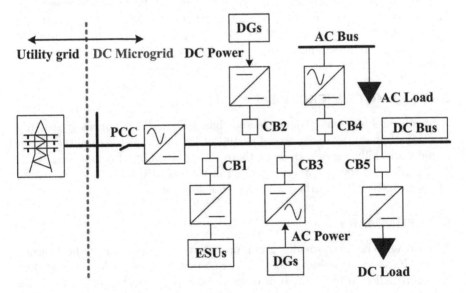

Figure 5.4 General structure of a DC microgrid

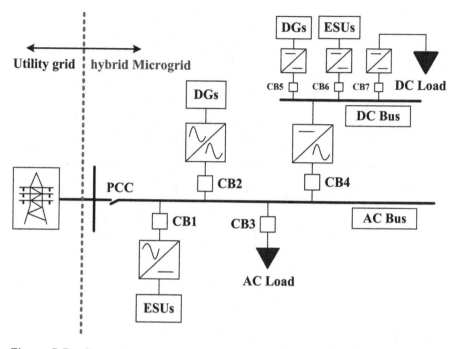

Figure 5.5　General structure of a hybrid microgrid

trade active and reactive powers with the utility grid by way of the PCC. This occurs when the MG is linked to the grid. In the event of abnormal operating conditions, an abnormal operating condition of any kind, or while performing maintenance on the utility grid, the urban MG can also be run in the islanded mode. These kinds of MG are appropriate for use in residential as well as commercial settings, such as hospitals, university campuses, shopping malls, and factories, among other types of establishments.

5.5.5　REMOTE MICROGRID

A remote MG is not as frequent as an urban MG, and it is purposely situated in areas that do not have access to the utility grid, such as military sites, islands, and hilly regions. Because there is no possibility of connecting to the utility grid, the remote MG is operated in the islanded mode only. A remote MG, in contrast to an urban MG, does not have any investment.

5.6　SUMMARY

The current chapter provides a comprehensive examination of the concept of an MG, encompassing a thorough exploration of its various definitions, afterward followed by an analysis of its key components. The multiple modes of functioning of the MG have been extensively investigated and examined. Subsequently, an illustrative MG configuration has been showcased to the readers, highlighting the notable importance of the MGCC. The most recent subject discussed is the classification of the MG based on both supply and location.

QUESTIONS

1. Give a detailed discussion on the concept of an MG along with providing various standard definitions.
2. What are the various entities that comprise an MG. Discuss in detail.
3. Discuss the various modes of operation of an MG in detail.
4. What are various challenges faced by an MG when operating in grid-connected mode as well as islanded mode?
5. Give a classification of the various types MGs along with their specifics.

6 Control Approaches for LFC and Metaheuristic Optimization Algorithms

Frequency is a key indicator of health of any power system. If the frequency profile is improved, the power system will be in better health, and vice versa. A persistent imbalance between the amount of power generated and the load demand can cause a permanent offset in the system frequency, which in turn causes expensive damage to the equipment that makes up the power system. It is expected that an effective control strategy will be implemented for the frequency regulation of the power system in order to produce a dependable, efficient, economically sound, and stable operation of the power system. In light of this, the following chapter provides a comprehensive overview of the many different control approaches that are used for the LFC study. At the outset, the significance of control approaches for LFC of the power system is depicted. In addition, a taxonomy of the numerous control approaches that are utilized most frequently is described one at a time, along with the benefits and constraints associated with each strategy. There is also a comparison and summary of all of the different control approaches presented here. In addition, the idea of metaheuristic optimization algorithms (MOA) is broken down in great detail, along with a general overview of their classification, the benefits they offer, and the restrictions they impose. Also, included at the end is a presentation of the mathematical formulation of a variety of performance indices that are routinely used.

6.1 WHY ARE CONTROL APPROACHES SIGNIFICANT FOR LFC?

Significance of the LFC in the power system has been previously discussed in section 1.2 of Chapter 1. Consequently, anyone with a vested interest can consult that particular section for further details. Ensuring the stability of frequency fluctuations is a critical task during the progressive expansion of the infrastructure of the power system. The demand is further compounded by the presence of parameter uncertainty and the diverse characteristics of the load. Frequency stabilization enhances the reliability, efficiency, and cost-effectiveness of power system operation, while also promoting its stability. Numerous approaches for the LFC have been proposed for implementation in the power system across several decades of extensive research and development. Currently, the power system is undergoing a shift from a vertically

DOI: 10.1201/9781003477136-6

integrated utility structure to a system that incorporates the DG, ES technologies, and PECs. As a result of this transformation, there is a need for more sophisticated and effective solutions for the LFC in order to ensure optimal operation and efficiency. A comprehensive LFC approach should possess the capability to effectively address the uncertainties arising from diverse DG sources, while also being equipped to handle the swift dynamics of the PECs.

The recent past has seen a rapid surge of the MGs owing to a number of advantages related to them as elucidated in the previous chapters. The integration of the RESs such as PV arrays and WTG unit in an MG poses a substantial challenge due to the inherent unpredictability and intermittent nature of these sources. This unpredictability and intermittent operation of the RES-based generators are fundamental concerns for the efficient operation of an MG. The absence of a direct link between the different DG sources and the MG bus, due to the existence of appropriate PECs, results in a notable reduction in the overall inertia of the MG. This effect is particularly pronounced when the MG is operating in the islanded mode. This is a matter that warrants recognition. Both of these variables possess the capacity to contribute to significant frequency deviation challenges in the MG whenever there is a modification in the load demand. Consequently, the reliability and stability of the MG may be jeopardized. Hence, the imperative arises for the advancement of a proficient LFC strategy that aims to restore the system frequency to its standard level. In the subsequent sections, an examination will be conducted on the various control approaches commonly employed for the LFC study of power systems.

6.2 CLASSIFICATION OF VARIOUS CONTROL APPROACHES FOR THE LFC

A general classification of the various control approaches available for the LFC is shown in Fig. 6.1. These are: (i) conventional control, (ii) robust control, (iii) cascade control, (iv) fractional-order control, and (v) degree-of-freedom control approaches. Fundamentals of each of these control approaches are discussed briefly in the following subsections.

6.2.1 CONVENTIONAL CONTROL APPROACH

Conventional control approach is the most widely implemented approach in industry as well as academia owing to its simple structure and ease of implementation. This approach is based on graphical design methods developed in the frequency domain like Bode diagram and root locus method. The proportional (P), integral (I), and derivative (D) controllers or combination of any of these are the commonly implemented controllers in this approach for the control of processes with feedback loop. Figure 6.2 shows a general scheme of a process with feedback loop and PID controller implemented in it. In the figure, $r(t)$ is the reference signal, $e(t)$ is the error signal, $u(t)$ is the PID

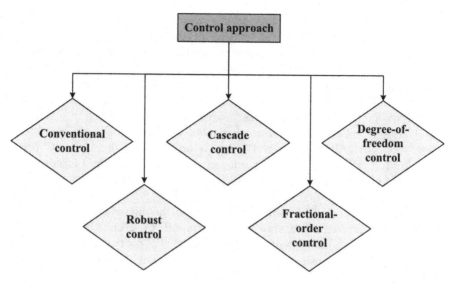

Figure 6.1 General classification of various control approaches for the LFC of power systems

controller output signal, and $y(t)$ is the process output signal. The controller signal is defined as

$$u(t) = Ke(t) + \frac{K}{t_i} \int_0^t e(\tau) \, d\tau + Kt_d \frac{de(t)}{dt} \qquad (6.1)$$

$Ke(t)$ represents the proportional term with gain K, $\frac{K}{t_i} \int_0^t e(\tau)d\tau$ represents the integral term with t_i as the integral time, and $Kt_d \frac{de(t)}{dt}$ represents the derivative term with t_d as the derivative time. Each term has its own significance: (i) the P term controls the transient specifications (like rise time, settling time, and over/undershoot) of the output response, (ii) the I term controls the steady state error of the output response, and (iii) the D term controls the damping of the output response.

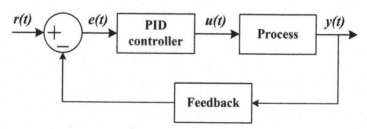

Figure 6.2 General scheme of a process with feedback loop implemented with PID controller

The previous studies extensively employed the conventional control approach to analyze the LFC study of the power system [90–97]. However, it should be noted that this approach is applicable only to a specific operational state of the power system when implemented. The approach indicated above exhibits a lack of agility in responding to dynamic operational conditions and non-linearities within the power system, hence indicating its inefficiency [98].

6.2.2 ROBUST CONTROL APPROACH

The robust control approach has a tendency for effectively minimizing fluctuations in system frequency and tie-line power, while also expanding the security margin to encompass all operational scenarios and potential configurations of the power system. Model predictive control (MPC), H_2/H_∞ control, μ theory, and internal model control (IMC) are the most extensively utilized robust control approaches.

6.2.2.1 Model predictive control approach

The MPC approach is well regarded in both industrial and academic domains. The MPC approach is employed to forecast the future behavior of a plant and afterward construct an explicit state-space model of the plant for control purposes. The approach has several benefits associated with its fundamental multi-variable formulation. These advantages encompass the incorporation of constraints, the ability to do online optimization, a straightforward design approach for complex systems, and the capability to fully compensate for delays in the system [3, 99]. Existing scholarly literature extensively explores a wide range of studies that have employed the MPC approach to analyze the LFC of power systems [13, 77, 100–105].

6.2.2.2 Mixed H_2/H_∞ approach

The mixed H_2/H_∞ approach integrates the advantageous characteristics of both the H_2 and the H_∞ synthesis procedures. Each of these synthesis procedures is insufficient to fully satisfy all the design objectives of a system. The design objectives encompass the aspects of disturbance rejection, stability, reference tracking, and constraint handling. During the implementation of the mixed H_2/H_∞ approach for a system, the H_2 synthesis is responsible for enhancing disturbance rejection and reference tracking capabilities, while the H_∞ synthesis focuses on handling constraints and maintaining system stability. Several works can be found in the literature utilizing the mixed H_2/H_∞ approach for the LFC analysis of the power system [106–110].

6.2.2.3 μ-theory approach

The μ-theory approach is founded on a comprehensive spectrum theory for matrices, which incorporates a structured singular value associated with the parameter μ to account for the structured uncertainty present in the system [111]. This design feature guarantees a resilient stability and optimal performance throughout a diverse set of system operating conditions, while also allowing for potential improvements in the feedback control capability. The efficacy of the μ-theory approach has been substantiated in several scholarly works of prior literature, particularly in the context of analyzing the LFC of the power systems [8, 112–115]. Interested readers can locate more details about the approach in [111, 116, 117].

6.2.2.4 Internal model control approach

The IMC approach is a resilient control approach that is characterized by its analytical nature, comprehensibility, and ability to handle sub-optimality and reduced computational complexity. The design approach of the approach closely resembles that of an open-loop controller. However, unlike the open-loop controller, the IMC approach exhibits enhanced effectiveness in addressing system uncertainties and disturbances [118]. Numerous referrals can be spotted in the literature employing the IMC approach for the LFC analysis of the power system [119–125].

Nevertheless, the robust control approach is practical for addressing multivariable circumstances and uncertainties. However, its design is complex and optimal performance can only be achieved when an accurate mathematical model of the power system is accessible [126].

6.2.3 CASCADE CONTROL APPROACH

The cascade control approach is another class of controllers that have been studied widely by the researchers. The cascade control approach possibly achieves fast rejection of the disturbance before it is distributed to other parts of the plant [127, 128]. The fact that a cascade controller possesses more tuning knobs as compared to non-cascade controller, improved performance can be expected from the former [129].

Figure 6.3 shows the implementation of a cascade control scheme in an appropriate typical control system. The transfer function of the system is

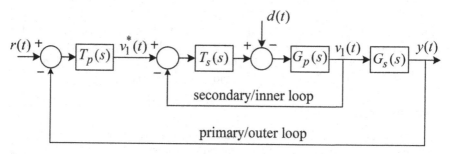

Figure 6.3 General scheme of a cascade control system

$G_p(s)G_s(s)$ and the system is subjected to a disturbance of $d(t)$. The cascade controller is implemented in two parts having transfer functions as $T_p(s)$ and $T_s(s)$. The inner loop consisting of $G_s(s)$ and $T_s(s)$ is called the secondary loop while the outer loop consisting of $G_p(s)$ and $T_p(s)$ is called the primary loop. The variable $v_1(t)$ between the $G_p(s)$ and $G_s(s)$ is measurable and $v_1^*(t)$ is the output signal from the primary controller that acts as a reference for the inner loop. In general, the outer loop controller contains an integral term in order to eliminate any steady-state errors in the output response of the system [130]. The inner loop controller can contain any of the combinations of the P, I, and D terms.

Following are the steps involved in the design of a cascade control scheme [130].

1. Initially segregate a complex control system into a number of first-order or second-order subsystems considering all the physical relationships and availability of measurements.
2. For each of the subsystems a controller is designed with any desirable combination of the P, I, and D terms as per the requirements. It is noteworthy here that first the inner loop is designed and its transfer function is obtained. Afterwards, considering the inner loop design the outer loop is planned.
3. Stability and performance analysis of the control system are conducted and closed-loop responses of both the inner loop and outer loop are appropriately adjusted. For a stable operation, the inner loop should have a much faster response as compared to the outer loop.
4. To achieve a fast response speed of the inner loop it is desirable to put a proportional control on the feedback error $(v_1^*(t) - v_1(t))$. This will also be useful in case the inner loop system is nonlinear.

Various researchers have successfully implemented the cascade control approach for the LFC analysis of the power system [72,73,131–135]. Nevertheless, the intricacy of a cascade control methodology gets complicated as a result of the inclusion of a supplementary controller. Concurrently, the process of tuning gets rigorous.

6.2.4 FRACTIONAL-ORDER CONTROL APPROACH

The fractional order (FO) calculus has a history spanning three centuries. However, it was not until the nineteenth century that these concepts achieved significant recognition and acceptance within the scientific and engineering communities. Since then, numerous academics have directed their efforts on studying this phenomenon with the aim of developing LFC controllers. In contrast to controllers that rely on integer order calculus, controllers based on the FO calculus offer increased flexibility in the design process due to the presence of additional tuning knobs. This enhanced flexibility allows for greater manipulation of the system dynamics [136,137].

FO calculus is a generalization of the integer order calculus that transforms the integer order integral or differential operators to fractional order. A fractional operator, denoted by $_AQ_t^{\bar{\alpha}}$, is utilized to carry out the transform. Sign of the operator $\bar{\alpha}$ decides the integration or differentiation operation. The fractional operator is modeled as [138]:

$$_AQ_t^{\bar{\alpha}} = \begin{cases} \frac{d^{\bar{\alpha}}}{dt^{\bar{\alpha}}}, & \bar{\alpha} > 1 \\ 1, & \bar{\alpha} = 0 \\ \int_{\bar{\alpha}}^{t}(dt)^{\bar{\alpha}}, & \bar{\alpha} < 1 \end{cases} \qquad (6.2)$$

Various definitions are available in the literature for describing the FO functions. Some commonly used definitions are the Riemann-Liouville (R-L) definition, the Grunwald-Letnikov (G-L) definition, and the Caputo definition [139]. Among these, the most frequently employed is the R-L definition, which is defined as

$$_AQ_t^{\bar{\alpha}} = \frac{1}{\Gamma(k - \bar{\alpha})} \frac{d^k}{dt^k} \int_{\bar{\alpha}}^{t} \frac{x(\tau)}{(t - \tau)^{1-(k-\bar{\alpha})}} d\tau \qquad (6.3)$$

where k is an integer, $k-1 < \bar{\alpha} \le k$ and $\Gamma(.)$ is Euler's gamma function. Before putting the FO controllers in simulation or practical applications, it is essential to approximate the FO integrators and differentiators with their integer order counterparts. Since the FO integrators and differentiators possess infinite dimensional nature, they must be first band-limited before putting into practice. In order to band-limit them certain approximation methods are available. These include the Carlson approximation, the Matsuda approximation, the Crone approximation and Oustaloup's recursive approximation [26, 140]. When using the Oustaloup's recursive approximation, which is most widely used, the FO integrator and differentiator $s^{\bar{\alpha}}$ is approximated through a recursive distribution of poles and zeros given by

$$s^{\bar{\alpha}} = M \prod_{k=1}^{K} \frac{1 + (s/\omega_{z,k})}{1 + (s/\omega_{p,k})} \qquad (6.4)$$

The frequencies of the zeros and poles are given by

$$\begin{cases} \omega_{z,1} = \omega_l\sqrt{\eta} \\ \omega_{z,k} = \omega_{z,k}\epsilon, \quad k = 1, 2, ..., K \\ \omega_{z,k+1} = \omega_{p,k}\sqrt{\eta}, \quad k = 1, 2, ..., K-1 \\ \epsilon = \left(\frac{\omega_u}{\omega_l}\right)^{\bar{\alpha}/K} \\ \eta = \left(\frac{\omega_u}{\omega_l}\right)^{(1-\bar{\alpha})/K} \end{cases} \qquad (6.5)$$

The FO controllers have been effortlessly administered for the LFC analysis of the power system by numerous researchers [26, 98, 126, 141–150]. However, the tuning process of an FO controller becomes clumsy due to the presence of additional tuning parameters.

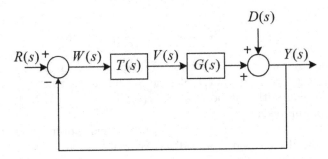

Figure 6.4   General scheme of a feedback control system with one DoF control

6.2.5   DEGREE-OF-FREEDOM CONTROL APPROACH

The degree-of-freedom (DoF) of a control system is defined as the number of closed-loop transfer functions that can be adjusted independently [151, 152]. Implementation of the DoF with a conventional controller enhances its set-point tracking and load disturbance rejection capabilities [153]. General scheme of a feedback control system with one DoF control is shown in Fig. 6.4. Here $G(s)$ represents the transfer function of the plant including the plant, actuator, and sensor dynamics, $T(s)$ represents the controller, $R(s)$ is the reference signal, $Y(s)$ is the plant output, $V(s)$ is the controller output signal, $W(s)$ is the error signal, and $D(s)$ is the disturbance.

Output of the plant is given by

$$Y(s) = \frac{G(s)T(s)}{1 + G(s)T(s)}R(s) + \frac{D(s)}{1 + G(s)T(s)} \tag{6.6}$$

The controller output signal is given by

$$V(s) = \frac{T(s)}{1 + G(s)T(s)}R(s) - \frac{T(s)}{1 + G(s)T(s)}D(s) \tag{6.7}$$

Assuming the $D(s)$ to be equal to zero, the ratios $\frac{Y(s)}{R(s)}$ and $\frac{V(s)}{R(s)}$ are given by

$$\frac{Y(s)}{R(s)} = \frac{G(s)T(s)}{1 + G(s)T(s)} \tag{6.8}$$

$$\frac{V(s)}{R(s)} = \frac{T(s)}{1 + G(s)T(s)} \tag{6.9}$$

In a similar manner, assuming the $R(s)$ to be equal to zero, the ratios $\frac{Y(s)}{D(s)}$ and $\frac{V(s)}{D(s)}$ are given by

$$\frac{Y(s)}{D(s)} = \frac{1}{1 + G(s)T(s)} \tag{6.10}$$

$$\frac{V(s)}{D(s)} = -\frac{T(s)}{1 + G(s)T(s)} \tag{6.11}$$

In the above four equations, selection of a suitable controller $T(s)$ makes all the transfer functions constant. Only one variable is left to change the output $Y(s)$ of the plant either it be $R(s)$ or $D(s)$. This makes it to be called one DoF control approach.

On the other hand, Fig. 6.5 shows a general scheme of implementation of a two DoF control approach in a feedback control scheme. As can be seen, the scheme consists of an extra block $F(s)$ connected after the reference signal $R(s)$. The plant output $Y(s)$ in this case is given by

$$Y(s) = \frac{G(s)T(s)R(s)}{1 + G(s)T(s)}R(s) + \frac{D(s)}{1 + G(s)T(s)} \tag{6.12}$$

From this equation, the ratios $\frac{Y(s)}{R(s)}$ and $\frac{Y(s)}{D(s)}$ are given by

$$\frac{Y(s)}{R(s)} = \frac{G(s)T(s)F(s)}{1 + G(s)T(s)} \tag{6.13}$$

$$\frac{Y(s)}{D(s)} = \frac{1}{1 + G(s)T(s)} \tag{6.14}$$

The component $F(s)$ provides one more DoF to influence the plant output $Y(s)$ along with the $R(s)$. This control approach thus forms a two DoF control approach.

Implementation of the two DoF-based controller for the LFC analysis of the power system can be located in several works in the literature [154–158].

In parallel, the three DoF controllers have also been implemented to obtain an optimal LFC performance of the power system [159–162]. The general scheme of a three DoF control approach integrated in a feedback control system is shown in Fig. 6.6 [161]. Output equation of the plant $Y(s)$ is also given

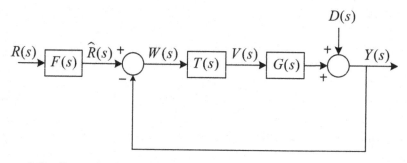

Figure 6.5 General scheme of a feedback control system with two DoF control

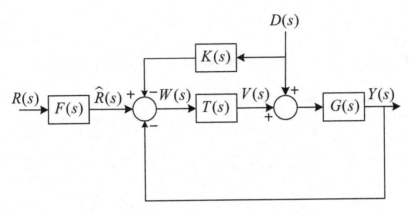

Figure 6.6   General scheme of a feedback control system with three DoF control

in (6.15). The component $K(s)$ forms the third DoF. However, the tuning process of a DoF-based controller also becomes clumsy due to the presence of additional tuning parameters.

$$Y(s) = \frac{G(s)T(s)F(s)}{1 + G(s)T(s)}R(s) + \frac{G(s) - G(s)T(s)K(s)}{1 + G(s)T(s)}D(s) \qquad (6.15)$$

6.3   SUMMARY OF ADVANTAGES AND LIMITATIONS OF VARIOUS CONTROL APPROACHES

Table 6.1 gives the advantages and limitations of the various control approaches presented.

6.4   METAHEURISTIC OPTIMIZATION ALGORITHMS

An MOA, also commonly and broadly known as an artificial intelligence algorithm, literally refers to an algorithm incorporating a metaheuristic implemented for the optimization of a process or a problem. The MOAs are employed for the purpose of seeking approximate solutions to optimization issues. Frequently, their utilization is employed in situations where the determination of an accurate solution proves to be excessively computationally burdensome. The MOAs operate through an iterative process wherein a solution is progressively enhanced until it reaches a satisfactory level of quality, hence attaining the status of a final solution. The term MOA is composed of three different terms: *metaheuristic* + *optimization* + *algorithm* that are addressed individually in the subsequent portion of this section.

The term *metaheuristic* combines two different words: *meta* which means 'beyond' and *heuristic* which means 'to find'. The metaheuristics belong to a class of approximate methods that were introduced in the early 1980s and

Table 6.1

Advantages and limitations of various control approaches

Control approach	Advantages	Limitations
Conventional control	1. simple in structure 2. ease of tuning	1. works inefficiently when operating conditions are varied
Robust control	1. works well under multivariable conditions 2. handles uncertainties effectively	1. complex in design 2. lack constraint handling capability 3. performs optimally only when exact mathematical model of the plant is known
Cascade control	1. achieves fast rejection of disturbance 2. presence of more tuning nubs improves system performance	1. requirement of an additional controller increases the controller complexity 2. tuning process becomes cumbersome
FO control	1. presence of extra tuning knobs provides more flexibility in design process 2. provides better chance to adjust the system dynamics	1. tuning process becomes clumsy due to presence of additional parameters
DoF control	1. enhances set-point tracking 2. improves disturbance rejection capability	1. tuning process becomes clumsy due to presence of additional parameters

since then have seen an unprecedented growth in their implementation. They find an upper edge when compared to the classical heuristics and optimization methods to solve complex optimization problems which becomes an obvious reason for their application. A metaheuristic is, in general, an iterative process that guides a supportive heuristic using an intelligent learning strategy to effectively explore and exploit the search space in order to achieve an optimal solution [163].

As per [164], the evolution of metaheuristics can be divided into five different stages: (i) *pre-theoretical period* (1940 and before), during which heuristics and even metaheuristics were used but not formally introduced, (ii) *early period* (1940 to 1980), during which the first formal studies on heuristics appeared, (iii) *method-centric period* (1980 to 2000), during which the field of metaheuristics truly took off and many different methods were proposed, (iv) *framework-centric period* (from 2000 till now), during which the insight grew that metaheuristics are more usefully described as frameworks, and not as methods, and (v) *scientific period* (the future), during which the design of metaheuristics would become a science rather than an art.

Optimization refers to the process of finding the best possible solution among all the possible solutions that tend to either minimize the operating cost of a system or maximize its efficiency. Simultaneously, the best possible solution has to satisfy all the system constraints. Consider the example of economic load dispatch problem of three thermal generators in any power system control area. Since the operating cost of each generator is a quadratic function of its real power output, the real power output of all the generators giving a minimum overall operating cost will correspond to an optimized economic load dispatch. The real power outputs of all three generators obtained for this optimized economic load dispatch will be considered as the best possible power outputs (optimal solution) of these generators.

Optimization-related studies can be located in the primitive history [165]. Around 300 BC, the famous Greek mathematician Euclid estimated the minimum distance of a point from a line. Apollonius of Perga, a Greek geometer, studied the greatest and least distances of a point from the perimeter of a conic section in around 200 BC. Pierre de Fermat, a French mathematician, in the 16th century and Joseph-Louis Lagrange, an Italian mathematician, in the 17th century, proposed calculus-based formulae for finding an optima. Carl Friedrich Gauss, a German mathematician, in the 18th century proposed iterative methods to search for an optimal solution. With the advent of the computer systems in the midst of the 19th century, it has been possible to optimize complex, multiobjective, multidisciplinary, and nonlinear problems.

We all are very much aware of the term algorithm that is commonly used in the field of mathematics and computer science. An algorithm is defined as [166]:

Any well-defined computational procedure that takes some value, or set of values, as input and produces some value, or set of values, as output

Say, for example, we are required to write a MATLAB program (in *.m* file) to add two matrices and determine the inverse of the resultant matrix. So the input to the program will be any two desired matrices. Afterwards, the two matrices will be added by using the + command and the inverse of the resultant matrix will be determined by using the *inv* command in the same file. And ultimately the output matrix will be displayed using the *disp* command. So as per the definition of an algorithm, we have entered two inputs (matrices) in the *.m* file, followed by a computational procedure to add and obtain the inverse of the resultant matrix and display it as the output.

History of an algorithm can be found in the Sumerian civilization which dates back to 2500 BC, the Babylonian civilization around 2000 BC, and the Egyptian civilization around 1550 BC that used to write algorithms in the form of certain symbols [167]. Now with the advancements in the computer technology, it is possible to write and simulate complex algorithms without consuming much time and efforts. An algorithm can be expressed using natural language, pseudocode, flowchart, and any programming language (just like the MATLAB example discussed above). Out of these, the pseudocode and the flowchart are the most preferable ways to express an algorithm.

While the MOAs have proven to be a useful approach for tackling optimization issues, it is vital to recognize that they are not without flaws. Individuals may be sensitive to the parameters used, making them vulnerable to specific affects. As a result, it is feasible that they will occasionally produce inferior results. Nonetheless, they continue to be a powerful tool that can be used to effectively address a wide range of complicated problems.

6.5   GENERAL STRUCTURE OF AN MOA

An MOA can be either a *single-solution search-based* or *population search based*. But here general structure of a population search-based MOA is being presented owing to its implementation for the LFC analysis to be discussed in the subsequent chapters. General structure of an MOA can be divided into three parts: (i) initialization, (ii) generation, comparison, and updation, and (iii) termination criterion as presented below:

---

initialization
% specifying the variables

- population size (*popsize*), dimension (*dim*), maximum number of iterations (t_{max}), lower bound (*lb*), upper bound (*ub*)

% initializing a set of potential solutions (x)

- $x(\text{init}) = x(lb) + rand(popsize, 1) \times (x(ub) - x(lb))$

% evaluating the cost function (F)

- evaluate the cost function (F) corresponding to each initialized value

generation, comparison, and updation
%start of iterations
for t = 1:t_{max}

- generate a new set of values (x) by using the initialized ones (x(init))

- compare the new values with the initialized ones in terms of minimized cost function (F(min))

- update the population with the x that correspond to comparative F(min)s

   for t = t+1

- generate another set of x by using those obtained during the t_{th} iteration

end

termination criterion

- repeat all the steps of the iterative process until the
  termination criterion is satisfied
- print the F(min) and corresponding values of x

6.6 CLASSIFICATION OF THE MOAs

A broad and general classification of the MOAs, as also highlighted in
the preceding section, is *single-solution search-based* and *population search
based* [168, 169]. A single-solution search-based MOA employs a single candidate solution and guides this solution to improve and simultaneously reach
an optimal solution in the search space. Examples of such MOAs include hill
climbing [170], simulated annealing [171], and tabu search [172]. On the other
hand, a population search-based MOA engages multiple candidate solutions
to explore the search space in order to ascertain an optimum solution. The
single-solution search-based MOAs often get stuck in a local optimum and
may not be able to locate a global optimum. Unlikely the population search-
based MOAs maintain diversity in the solutions with lesser chances of getting
stuck in a local optimum and thus result in an enhanced exploration of the
search space.

The population search-based MOAs can further be classified into the
following types: (i) evolution-based MOAs, (ii) physics-based MOAs, and
(iii) swarm-based MOAs [202]. The evolution-based algorithms implement a
population-based strategy and diversity in the population is adhered by mim-
icking certain genetic rules that include reproduction, elimination, mutation,
selection, chemotaxis, and migration during the course of iterations. In these
MOAs, the individuals in the population are exposed to some environmental
pressure as a result of which only the fittest individuals survive and hence
average fitness of the population is increased.

The physics-based MOAs are inspired from the physical laws of the uni-
verse. Proposal of the quantum computing system almost four decades ago
resulted in the emergence of the physics-based MOAs [203].

The swarm-based MOAs mimic the behavior of self-organization and shape-
formation of swarms in nature. Self-organization and decentralized control are
the principal features of the swarm-based MOAs that lead to their emergent
behavior. This particular behavior of these algorithms is possible only due to
interactions between individual members in the population and is not possible
to achieve when the individual members act alone [204]. Elite members of the
swarm are given the preference during the course of the iterations. A pictorial
presentation of all the above classifications is shown in Fig. 6.7.

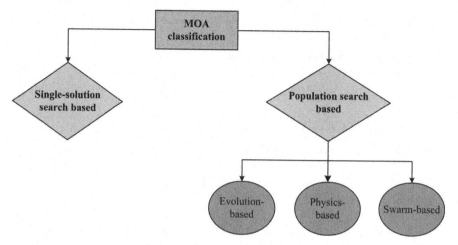

Figure 6.7   Classification of the MOAs

Table 6.2 shows the various commonly used algorithms relevant to the classification of the population search-based MOAs.

6.7   APPLICATION OF THE MOAs IN POWER SYSTEM

A large number of generation, transmission, and distribution subsystems comprise a power system. Due to the size and complexity of the system, its installation and administration are extremely challenging. In essence, an electrical system is a very large network containing very large data collections. The analysis and resulting implementation of these data sets can take considerable time. An electrical power system is essential, but if improperly operated and managed, it can be extremely hazardous. The demand for electricity is ever-increasing, posing challenges and difficulties in maintaining load demand without overwhelming the system.

For the planning, installation, operation, and control of such a large system, it is necessary to employ modern technologies. Applications employing the MOAs have a number of essential capabilities that can aid in supporting a power system and managing power system operations overall. MOA-based applications are able to manage the enormous data sets associated with a power system. Moreover, they can aid in the design of power plants, the modeling of installation layouts, the optimization of load dispatch, and the rapid response of control apparatus. These applications and their methodologies have met with tremendous success in a variety of sub-fields of power system engineering.

Although this book is specifically attributed to MOA-based LFC study of the MGs, there are several other power system optimization issues where the MOAs have been implemented. Some of the key optimization issues include fault detection and diagnosis [205–207], power quality enhancement [208–210],

Table 6.2

Some commonly used population search based MOAs

Evolution-based MOAs	Differential evolution [173] Genetic algorithm [174] Genetic programming [175] Evolution strategy [176] Evolution programming [177] Biogeography-based optimization [178] Probability based incremental learning [179] Invasive weed optimization [180] Grammatical evolution [181]
Swarm-based MOAs	Salp swarm algorithm [182] Krill herd algorithm [183] Ant colony optimization [184] Firefly algorithm [185] Hunting search [186] Social spider algorithm [187] Bird mating optimizer [188] Moth swarm algorithm [189] Dolphin echolocation algorithm [190] Migrating birds optimization [191]
Physics-based MOAs	Gravitational search algorithm [192] Vortex search algorithm [193] Galaxy-based search algorithm [194] Binary magnetic optimization algorithm [195] Ion motion algorithm [196] Water wave optimization [197] Teaching-learning-based optimization [198] Thermal exchange optimization [199] Heat transfer search algorithm [200] Water cycle algorithm [201]

energy management [211–213], generation scheduling [214–216], economic load dispatch [217–219], and power system protection [220–222]. Figure 6.8 shows a pictorial representation of these issues.

6.8 ADVANTAGES AND LIMITATIONS OF THE MOAs

The MOAs, since the last two decades, have become very much popular among the research community and are consistently being implemented for solving various real-world optimization problems related to the fields of engineering, economics, and sciences. In general, the principal advantages of employing an MOA are [202, 223]:

1. *Applicability to various types of multidisciplinary problems:* The MOAs are flexible and can be applied to a variety of optimization problems, such

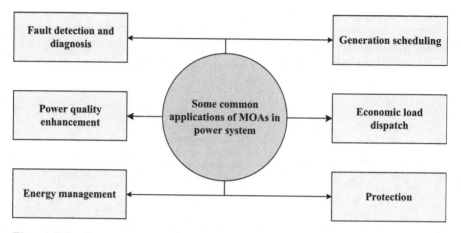

Figure 6.8   Some common applications of the MOAs in power system

as combinatorial, continuous, mixed-integer, and multi-objective problems. This adaptability makes them appropriate for an extensive range of practical applications.

2. *Global search:* The MOAs are able to conduct a global search of the solution space, allowing them to investigate a vast array of potential solutions. This aids in locating global or near-global optima, even in complex, multimodal, or non-convex landscapes of optimization.

3. *Easy of implementation:* The MOAs are frequently relatively simple to implement, particularly in comparison to some mathematical programming techniques. This accessibility makes them approachable to a broad spectrum of practitioners and users.

4. *Derivative-free operating mechanism:* The MOAs are gradient-free, unlike traditional optimization algorithms that rely on derivatives (e.g., gradient-based methods). This allows them to be utilized in situations where derivatives are either unavailable or computationally costly to compute.

5. *Avoidance from being getting trapped in a local optima:* In general, the MOAs are effective at avoiding local optima. In order to explore new regions of the solution space, algorithms such as genetic algorithms and simulated annealing use probabilistic mechanisms to take worse solutions with a certain probability.

6. *Robustness:* In general, the MOAs are robust to chaotic or imprecise objective functions. They can accommodate uncertain evaluations of solutions and are applicable to real-world problems where objective functions may contain inherent uncertainty.

However, each of the MOA presented in Table 6.2 has its own advantages and limitations. For any MOA to perform adequately, it is obligatory to maintain a good balance between the exploration and exploitation during the course of iterations [164]. Holding a proper balance between the exploration and exploitation depends upon a proper selection of the controlling parameter of an MOA. Choosing an inappropriate value of the controlling parameter may lead to perform the MOA in an unexpected manner thus deviating it away from finding the global optima. Table 6.3, at the end of this chapter, presents the specific advantages and limitations of some commonly implemented MOAs [224].

6.9   MATHEMATICAL FORMULATION OF PERFORMANCE INDICES

A performance index is utilized to quantitatively evaluate the performance level of any simulated control system. Any control system must incorporate a performance index in order for its operation to be adaptive, for its control parameters to be automatically optimized, and for its design to be optimal. As defined in [225], a performance index is

A quantitative measure of the performance of a system and is chosen so that emphasis is given to the important system specifications

Any control system is considered to be optimally designed if its performance index achieves an extreme value, in general, a minimum value. A useful performance index is the one having either a positive or zero value [225]. In modern control theory, generally four integral performance indices are most commonly used. These are: (i) integral of squared error (ISE), (ii) integral of absolute error (IAE), (iii) integral of time multiplied squared error ($ITSE$), and (iv) integral of time multiplied absolute error ($ITAE$). Each index is briefed below.

6.9.1   INTEGRAL OF SQUARED ERROR (ISE)

The ISE index evaluates the system performance by integrating the squared error over a definite time interval. The ISE index possesses smaller over/undershoots although results in a large settling time. It penalizes large errors more than small errors. This index is formulated as

$$ISE = \int_{0}^{tsim} e(t)^2 \, dt \qquad (6.16)$$

where, $e(t) = r(t) - y(t)$

Table 6.3

Advantages and limitations of some commonly implemented MOAs

MOA	Advantage(s)	Limitation(s)
Genetic algorithm	1. easy to implement 2. can effectively handle random types of objective functions and constraints (including nonlinear or discontinuous) 3. independent of other heuristics, can freely solve any given problem 4. employ simple operators and can be used to solve problems with high computational complexity	1. vulnerable to getting trapped in local optima, hence do not guarantee to reach a global optima 2. greater chances of occurrence of premature convergence, this results in loss of population diversity 3. becomes time-consuming if the number of variables are increased 4. absence of proper termination criterion, proper method of parameter setting
Differential evolution	1. possess good exploration capabilities 2. easy to implement, depends only upon a few parameters 3. can efficiently handle nonlinear, non-differentiable and multi-modal objective functions 4. can effectively handle cost functions with high computational complexity	1. gets easily trapped in local optima 2. does not possess a stable convergence 3. using the same parameter setting may lead to different global optima 4. requires repeated parameter setting
Particle swarm optimization	1. calculations are straightforward, suitable for dynamic systems 2. very much popular in scientific and engineering communities	1. suffers from partial optimism 2. all solutions tend to converge prematurely and consequently population diversity is lost
Firefly algorithm	1. population gets automatically divided into groups, hence exploration capabilities are enhanced	1. enhanced exploration capability can lead to reduced speed and degraded convergence rate 2. efficient performance depends upon proper parameter settings 3. not suitable for handling complex optimization problems

Algorithm	Advantages	Disadvantages
Invasive weed optimization	1. all potential solutions are involved in reproduction unlike GA where only the solutions with less fitness values are only involved 2. computational cost is nominal	1. exploration of large search space requires large number of seeds, hence the whole process becomes time consuming
Flower pollination algorithm	1. easy to implement, flexible, few parameter dependent 2. can efficiently handle single and multi-objective optimization problems	1. convergence rate is slower, low precision 2. gets easily trapped in a local optima
Cuckoo search	1. can efficiently discover true global optima in search space 2. can effectively balance between exploration and exploitation	1. produces low classifications accuracy 2. possess poor convergence rate
Gravitational search algorithm	1. dependence on only two parameters 2. possess good exploration capabilities to find a near-optimal solution	1. susceptible to getting trapped in a local optima 2. using the same parameter setting may lead to different global optima

$e(t)$ is the error, $r(t)$ is the reference input to the system, $y(t)$ is the output response of the system, $tsim$ is some finite simulation time.

6.9.2   INTEGRAL OF ABSOLUTE ERROR (IAE)

This performance index integrates the absolute value of error of a finite time interval. This index is mainly employed in the digital simulation of a system, although for real-time analytical analysis it is not considered to be feasible owing to computational complexity. Furthermore, the application of the IAE produces a slower system response. The IAE index is formulated as

$$IAE = \int_0^{tsim} |e(t)|\, dt \tag{6.17}$$

6.9.3   INTEGRAL OF TIME MULTIPLIED ABSOLUTE ERROR ($ITAE$)

The $ITAE$ index exhibits an additional time multiplier over the whole integration process that results in an improved system response as compared to the ISE and the IAE indices. Implementation of the $ITAE$ index offers lower over/undershoots as well as reduces the settling time of the system response. The $ITAE$ index is formulated as

$$ITAE = \int_0^{tsim} t|e(t)|\, dt \tag{6.18}$$

6.9.4   INTEGRAL OF TIME MULTIPLIED SQUARED ERROR ($ITSE$)

The $ITSE$ index also exhibits an additional time multiplier over the whole integration process that results in an improved system response. Furthermore, it combines the advantage of both the ISE and the $ITAE$ criteria thereby minimizing the over/undershoots and settling time of the system output response. The $ITSE$ index is formulated as

$$ITSE = \int_0^{tsim} t(e(t))^2\, dt \tag{6.19}$$

Figure 6.8 shows the plots of $r(t)$, $y(t)$, and $e(t)$ for a second-order system with unity feedback. Figure 6.9 illustrates the variations of the performance indices that have been previously described. The plot illustrates that the $ITSE$ index possesses the lowest value among all the indices.

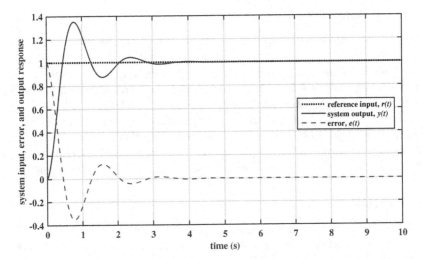

Figure 6.9   Reference input and variation of the system output and error signal for a second-order system

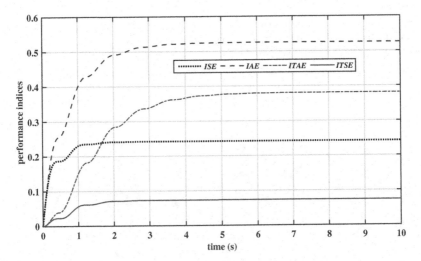

Figure 6.10   Variation of various performance indices

6.10 SUMMARY

This chapter explores the significance of the control approaches for the LFC of the conventional power system, which are followed by the rapidly advancing MGs. Several commonly used control approaches for LFC analysis are discussed in depth. These include the conventional control approach, the robust control approach, the cascade control approach, the fractional order control approach, and the degree-of-freedom approach. Further, a summary of the benefits and drawbacks of each of these control methods is provided. Finally, mathematical formulation of the various performance indices commonly implemented to measure the performance of a control system is presented.

QUESTIONS

1. What is the significance of control approaches for the LFC? Discuss.
2. Give a detailed classification of the various control approaches available for the LFC along with their advantages and limitations.
3. What is an MOA? Discuss with suitable examples.
4. Discuss the general structure of an MOA along with its principal classifications.
5. What are the key applications of an MOA from power system point of view. Discuss.
6. Give a detailed classification of the various performance indices commonly used along with mathematical formulation of each.

7 LFC Study of an Islanded Microgrid

This chapter examines the LFC efficacy of an islanded MG. A comprehensive dynamic model of the isolated MG is introduced at the outset of the discussion. In addition, a mathematical model of a PI-PD cascade controller constructed as the secondary controller in the feedback path of the MG is presented. During this procedure, the salp swarm algorithm (SSA) is used to simultaneously canvass the steps required to optimize the parameters of the cascade controller. In conclusion, a thorough analysis and discussion of the time-domain simulation results for the evaluated isolated MG is presented. This is followed by an analysis of the sensitivity and stability of the controller.

7.1 INTRODUCTION

As discussed in Chapter 5 of this book, when an MG is operating in islanded mode, it is accountable for independently managing the V/f and P/Q controls. In addition, a portion of the total number of connected DG sources assume the role of master DG sources, while the remainder assume the role of slave DG sources. Because the electrical power supplied by the DG sources is relatively low compared to the power generated by the utility grid, the loads connected to the MG in such a system are prioritized to ensure that the critical loads always have access to an uninterrupted power supply. In this mode, the operation of the MG becomes challenging due to the presence of RES-based DG sources with fluctuating power output and a low inertia of the MG as a result of the lack of direct connection between the various DG sources and the bus. These factors make the operation of the MG more challenging. The operation of the MG in islanded mode presents additional challenges, such as those associated with power quality, power sharing, and communication between the numerous DG sources that have been deployed.

This chapter comprehensively examines the LFC performance of an islanded MG whereby the dynamic model of the islanded MG is presented along with the mathematical modeling of the SSA-based PI-PD cascade controller and discussion of the time-domain simulation results in the following sections.

7.2 DYNAMIC MODEL OF ISLANDED MG

Figure 7.1 displays the dynamic model (frequency response model) of the islanded MG that is to be examined [17,84]. Blocks of first-order linear transfer functions of the various DG sources and ESUs are included in its model.

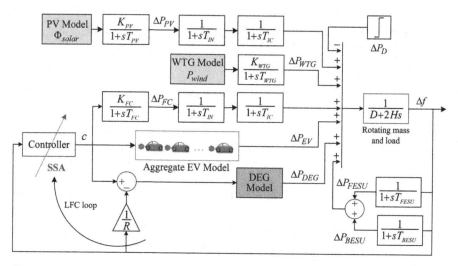

Figure 7.1 Dynamic model of the islanded MG

Since the LFC investigation is focused on small-signal analysis, the first-order linear transfer function blocks are sufficient [8, 23, 24]. On the demand side, all the DG sources, with the exception of those based on the RESs, function as a continuous source of high-quality power to match the generation and load demand. Unfortunately, due to their poor response times, they are aided by the various ESUs that have relatively long response times in order to compensate for any rapid shift in load [77, 126]. The frequency deviation signal, denoted by Δf, is received by the controller that is implemented in the LFC loop. In response to this signal, the controller modulates the output powers of the DEG, FC, and EV units. Due to the fact that they are dependent on the weather in order to function, the RES-based units, such as the WTG and the solar PV, are not included in the LFC loop.

Modeling of the various subsystems (units) comprising the islanded MG is discussed below:

7.2.1 WTG MODEL

A WTG unit being RES-based is variable and unpredictable in its nature. Consequently, this directly impacts the power output of the WTG unit making it highly fluctuating. Power output of the WTG unit is given by [19, 96]:

$$P_{wind} = \frac{1}{2}\rho A_r C_p V_w^3 \tag{7.1}$$

where
ρ: denotes the air density (kg/m^3)
A_r: denotes the area swept (m^2)

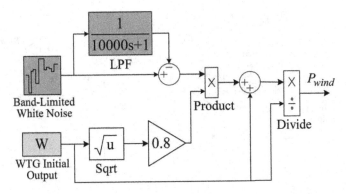

Figure 7.2 Transfer function model of a practical WTG unit

C_p: denotes the power coefficient that varies between 0.2 and 0.5
V_w: denotes the wind velocity (m/s)
Although the WTG unit is a highly non-linear system, the current LFC analysis, being related to small-signal analysis, utilizes a linearized model for the same whose transfer function is given by [19]:

$$G_{WTG}(s) = \frac{\Delta P_{WTG}}{P_{wind}} = \frac{K_{WTG}}{1 + sT_{WTG}} \qquad (7.2)$$

Transfer function model of a widely implemented practical WTG unit is shown in Fig. 7.2 [97, 226]. The model utilizes a band-limited white noise block with a low pass filter (LPF) considering elimination of components above 10000 s. The resulting output is multiplied by standard deviation, i.e., $0.8\sqrt{u}$ in order to derive a random power output (P_{wind}) from the WTG unit. W is the initial output of the WTG unit.

7.2.2 PV MODEL

A solar PV unit is also RES-based that makes its operation again variable and unpredictable. The power output of a solar PV unit is given by [19, 96]:

$$P_{PV} = \eta \phi_{solar} S [1 - 0.005 (T_a + 25)] \qquad (7.3)$$

where
η: denotes the conversion efficiency (%) of the PV module ranging from 9% to 12%
ϕ_{solar}: denotes the solar irradiation measured ($Watt/m^2$)
S: denotes the measured surface area of the PV unit (m^2)
T_a: denotes the ambient temperature (^oC)
Interfacing of the PV unit with the AC bus of the islanded MG is done via an inverter and an interconnector as shown in Fig. 7.1. This is so because a PV unit generates DC power. Although a PV unit is also a non-linear system but

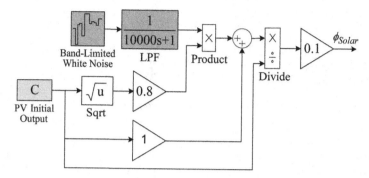

Figure 7.3 Transfer function model of a practical PV unit

for the present LFC study, it is represented by a first-order transfer function (as is with the WTG unit) given by [19]:

$$G_{PV}(s) = \frac{\Delta P_{PV}}{\phi_{solar}} = \frac{K_{PV}}{1 + sT_{PV}} \tag{7.4}$$

Transfer function model of a practical solar PV unit is shown in Fig. 7.3 [97]. The band-limited white noise and LPF blocks shown will work in the same manner as in the case of the WTG model of Fig. 8.2. Afterward, the resulting output is multiplied by standard deviation to generate a random power from the PV unit (ϕ_{solar}). C denotes the initial PV power output.

7.2.3 EV MODEL

As discussed in Chapter 4, the EVs are considered to be the future of the automotive industry owing their various advantages as already highlighted. Also, to participate in the frequency control process of a utility grid, an EV has to operate in the V2G mode of interaction depending on its battery SoC. The EVs with their battery SoCs (SoC_{avg}) in the range of 85 ± 5% are considered to participate in the V2G mode here. Lower limit of the SoC (SoC_{min}) is selected as 80% and upper limit of the SoC (SoC_{max}) is kept at 90% respecting the users' convenience during their next trip and life of the battery of the EVs [12, 70]. It is assumed that the EVs with 100% SoC are not controlled and hence cannot participate in the V2G mode of interaction.

An aggregate EV model developed in [10, 12, 14] is considered as shown in Fig. 7.4. The model is developed assuming that the SoC of all the EVs is based upon the synchronized SoC control method as presented in chapter 4. In the figure, input to the model is the LFC signal sent from the central load dispatching center in order to ensure a synchronized SoC control of all the participating EVs. The total charging/discharging power output of the EVs is denoted by ΔP_{EV}. $E_{control}(t)$ is the total energy of all the controllable EVs which is the output of the total energy model (TEM). After charging, the

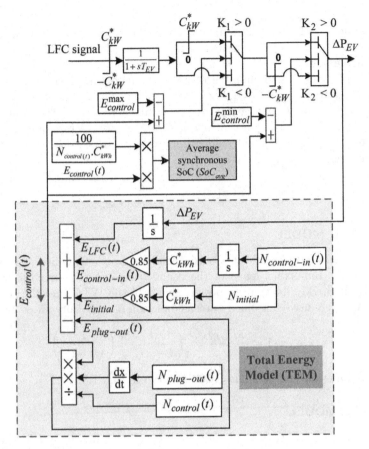

Figure 7.4 Aggregate EV model

EVs can respond to the LFC signal only when the condition

$$E_{control}^{min} \leq E_{control}(t) \leq E_{control}^{max} \qquad (7.5)$$

holds true. Here, $E_{control}^{min}$ and $E_{control}^{max}$ are the minimum and the maximum energy capacity limits of the EV, respectively. Their values are determined by

$$E_{control}^{min} = \frac{N_{control}(t).C_{kWh}^*}{1000} \times 0.8 \qquad (7.6)$$

$$E_{control}^{max} = \frac{N_{control}(t).C_{kWh}^*}{1000} \times 0.9 \qquad (7.7)$$

$N_{control}(t)$ is the total number of controllable EVs determined by

$$N_{control}(t) = N_{initial} + N_{control-in}(t) - N_{plug-out}(t) \qquad (7.8)$$

where

$N_{initial}$: denotes the initial number of controllable EVs

$N_{control-in}(t)$: denotes the number of EVs transferring from charging state to controllable state

$N_{plug-out}(t)$: denotes the number of EVs transferring from controllable state to driving state

The TEM model that generates the total energy of all the controllable EVs ($E_{control}(t)$) is also shown in Fig. 7.4. For more details about the aggregate EV model interested readers may refer to [10, 12, 14]. The SoC_{avg} is mathematically formulated as

$$SoC_{avg}(t) = 100 \times \frac{E_{control}(t)}{N_{control}(t).C_{kWh}^*} \ \%\qquad(7.9)$$

7.2.4 DEG MODEL

A conventional DEG facilitates a continuous supply of quality and reliable power to the important loads in the islanded MG. This is mandatory considering the highly fluctuating power output of the WTG and the solar PV units. A widely used simplified transfer function of it is represented by [19]:

$$G_{DEG}(s) = \frac{\Delta P_{DEG}}{c} = \left(\frac{1}{1+sT_G}\right)\left(\frac{1}{1+sT_T}\right)\qquad(7.10)$$

Transfer function model of the DEG is shown in Fig. 7.5 [14, 17].

7.2.5 FC MODEL

Key features of an FC have already been dealt with in Chapter 3 regarding its conversion of chemical energy into electrical energy. Like the conventional DEG unit, an FC unit is dedicated to compensate the imbalance between generation and load demand arising from the fluctuating behavior of the WTG and the solar PV units and thus maintain a stable operation of the system.

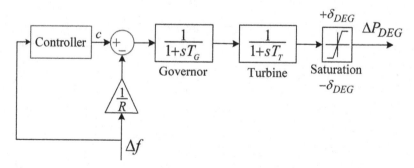

Figure 7.5 Transfer function model of a conventional DEG unit

To interface the FC block in Fig. 8.1 with the AC bus of the MG, an inverter and an interconnector are utilized. The first-order transfer function of an FC is given by [19]

$$G_{FC}(s) = \frac{\Delta P_{FC}}{c} = \frac{K_{FC}}{1 + sT_{FC}} \tag{7.11}$$

7.2.6 ESU MODEL

A myriad of ESUs have been discussed in Chapter 3 that comes in different forms be it physical, electromagnetic, or electrochemical. The battery energy storage unit (BESU) and the flywheel energy storage unit (FESU) are the most widely implemented ESUs (BESU being of electrochemical type and FESU being of physical type) for pacifying the frequency deviations in an MG. Principal reason for utilizing the ESUs in an MG is that their response time is faster as compared to the various DG units. This fact provides them the ability to readily compensate for any mismatch between the generation and load demand thus ensuring the reliability and stability of the MG. The ESUs rapidly charge/discharge when the generation exceeds the load demand and when the load demand exceeds the generation, respectively. The first-order transfer functions of the BESU and the FESU are given by [19]

$$G_{BESU}(s) = \frac{\Delta P_{BESU}}{\Delta f} = \frac{1}{1 + sT_{BESU}} \tag{7.12}$$

$$G_{FESU}(s) = \frac{\Delta P_{FESU}}{\Delta f} = \frac{1}{1 + sT_{FESU}} \tag{7.13}$$

Parameter values of the islanded MG are given in **Appendix A**.

7.3 CONTROLLER MODEL AND OPTIMIZATION OF ITS PARAMETERS

This section solicits the mathematical model of the PI-PD cascade controller implemented for the LFC analysis of the islanded MG. Further, since the performance of a controller depends upon the proper optimization of its tuning parameters, an optimizing algorithm (SSA) and its implementation for optimization of the controller parameters are discussed.

7.3.1 MODELING OF THE CONTROLLER

The investigated islanded MG shown in Fig. 7.1 utilizes the PI-PD cascade controller in its feedback path as the secondary controller. Input to the controller is the Δf. In accordance with the input signal received, the cascade controller regulates the power outputs of the EV, DEG, and FC in order to stabilize the deviations in the system frequency. The specifics, advantages, and limitations of a cascade control approach have already been dealt with in

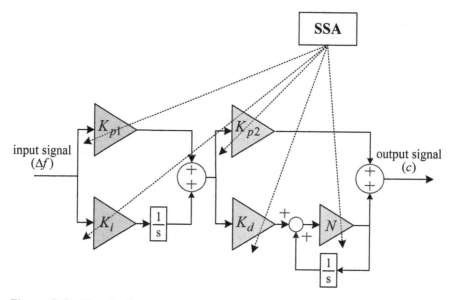

Figure 7.6   Tranfer function model of the PI-PD cascade controller

chapter 6. In the implemented PI-PD cascade controller, the PI controller acts as the primary controller and the PD controller with derivative filter (N) acts as the secondary controller. Transfer function model of the cascade controller is shown in Fig. 7.6. Transfer functions of the primary and the secondary controllers are given by

$$T_p(s) = K_{p1} + \frac{K_i}{s} \tag{7.14}$$

$$T_s(s) = K_{p2} + K_d \left(\frac{N}{1 + N/s} \right) \tag{7.15}$$

where, K_{p1} and K_i are the proportional and integral gains, respectively of the primary controller, K_{p2}, K_d, and N are the proportional gain, integral gain, and derivative filter coefficient, respectively of the secondary controller.

7.3.2   OPTIMIZATION OF THE CONTROLLER PARAMETERS

It is a common consent that optimal performance of any controller is directly related to an appropriate optimization of its tuning parameters. In this regard, the SSA has been utilized. The algorithm was proposed by Mirjalili *et al.* in the year 2017 [182]. The algorithm is simple, very easy to implement, and is hardly any parameter specific. Inspiration for the SSA comes from the swarming behavior of salps that reside in deep oceans searching for food. A salp has a barrel-shaped transparent body just like a jellyfish through which it propels water to move forward. The salps often navigate the oceans in a swarm referred to as the salp chain. The first member of the salp chain is referred to as the leader and the rest of the salps in the chain are the followers. Further details about the SSA can be located in [182].

An ISE-based objective function (OF_{ISE}) is utilized in order to optimize the parameters of the cascade controller. The LFC problem is thus formulated as:

Minimize:

$$OF_{ISE} = \int_0^{t_{sim}} \Delta f^2 dt \qquad (7.16)$$

Subject to:

$$\begin{cases} K_{p1min} \le K_{p1} \le K_{p1max} \\ K_{imin} \le K_i \le K_{imax} \\ K_{p2min} \le K_{p2} \le K_{p2max} \\ K_{dmin} \le K_d \le K_{dmax} \\ N_{min} \le N \le N_{max} \end{cases} \qquad (7.17)$$

where $K_{(p1,p2,i,d)min}$ and $K_{(p1,p2,i,d)max}$ are the lower bound (lb) and upper bound (ub) of the PI-PD cascade controller parameters, respectively. N_{min} and N_{max} are the lb and ub, respectively of the derivative filter coefficient N of the controller. Range for the parameters K_{p1}, K_i, K_{p2}, and K_d is considered in [-5, 5] and for N the range is [40, 120] after performing several trial runs. t_{sim} denotes the simulation time in seconds (s). Below are the steps to optimize the cascade controller parameters using the SSA.

```
% Optimization of the PI-PD cascade controller parameters
using the SSA

% specifying the inputs
```
- $popsize$ = 30; dim = 5; t_{max} = 100; N_{min} = 40; N_{max} = 120;
- $K_{min(p1,p2,i,d)}$ = -5; $K_{max(p1,p2,i,d)}$ = 5;

```
% initializing a population (set) of search agents (salps)
```
- $K_{p1} = K_{minp1} + rand(popsize, 1) \times (K_{maxp1} - K_{minp1})$;
- $K_{p2} = K_{minp2} + rand(popsize, 1) \times (K_{maxp2} - K_{minp2})$;
- $K_i = K_{mini} + rand(popsize, 1) \times (K_{maxi} - K_{mini})$;
- $K_d = K_{mind} + rand(popsize, 1) \times (K_{maxd} - K_{mind})$;
- $N = N_{min} + rand(popsize, 1) \times (N_{max} - N_{min})$;

```
% evaluating the objective function value
```

- evaluate the objective function (OF_{ISE}) corresponding to each initialized set of search agents using (7.16)

% start of iterations
for t = 1:t_{max}

- update the SSA parameter r_1 using
$r_1 = 2e^{-\frac{4t}{t_{max}}}$

- for each salp (s_i)
if i = 1
update the position of the leading salp using

$$s_j^1 = \begin{cases} Q_j + r_1\left((ub_j - lb_j)\,r_2 + lb_j\right), & if \ r_3 \geq 0 \\ Q_j - r_1((ub_j - lb_j)r_2 + lb_j), & if \ r_3 < 0 \end{cases}$$

else
update the position of the follower salp using

$$s_j^i = \frac{1}{2}\left(s_j^i - s_j^{i-1}\right)$$

end
end

- modify the position of the salps based on lb and ub

end

% termination criterion

- repeat all the steps till t reaches t_{max}
- print the salp with minimum value of the OF_{ISE} and corresponding values of the K_{p1}, K_{p2}, K_i, K_d, and N

Here, s is a two-dimensional matrix to store the positions of all the salps, s_j^1 denotes the position of the leader salp in the jth dimension, Q_j denotes the position of the food source in the jth dimension, ub_j and lb_j are the upper and lower bounds of the jth dimension, respectively. r_1, r_2, and r_3 are the random numbers. r_1 is of utmost importance for the SSA as it maintains a balance between exploration and exploitation capabilities of the algorithm. r_2 and r_3 are uniformly distributed in [0,1]. Key features of the SSA can be summarized as below [182]:

- Position of only the leader salp is updated w.r.t. the food source, hence the leader salp is responsible for exploring and exploiting the entire search space.
- Position of the follower salps is updated w.r.t. each other only and they gradually follow the leader salp during the entire course of iterations.

- Gradual movement of the follower salps prevents the algorithm from being getting trapped in a local optima.
- During the course of the iterative process, the parameter r_1 decreases gradually in order to maintain a balance between exploration and exploitation.
- The parameter r_1 is the only one controlling parameter of the SSA.

7.4 SIMULATION RESULTS

Dynamic responses (FDRs) of the islanded MG are presented and discussed in this section. To establish the potency of the PI-PD cascade controller its performance is compared with the conventional PID controller considering three different load disturbance conditions in the islanded MG. Comparative FDRs are presented for all the conditions in this regard. Additionally, the effect of the EVs on the LFC performance of the MG is demonstrated for all the considered load disturbance conditions. All the simulations were carried out in MATLAB R2015a installed on a personal laptop. The robustness of the controller against the parametric variations in the MG is also validated by performing the sensitivity analysis. Further, stability analysis is discussed in terms of root locus plots corresponding to all the load disturbance conditions.

The optimized values of the controller parameters correspond to $K_{p1} = 1.82$, $K_{p2} = -3.39$, $K_i = 2.71$, $K_d = -2.78$, and $N = 49.21$. It is to be noted here that these values have been kept the same when considering all the load disturbance conditions in the islanded MG. Further, in this LFC analysis, only EVs with a battery SoC between 80 and 90% are considered with $N_{control}(t) = 3600$, $N_{initial} = 5000$, $N_{control-in}(t) = 1100$, and $N_{plug-out}(t) = 2500$.

7.4.1 COMPARATIVE FDRs OF THE ISLANDED MG

Figure 7.7 shows the comparative FDRs of the islanded MG when considering a 1% step load perturbation (SLP). Variation in the solar PV power output (ΔP_{PV}) and WTG power output (ΔP_{WTG}) is not considered and these are kept constant. The impact of integrating the EVs on the FDRs is

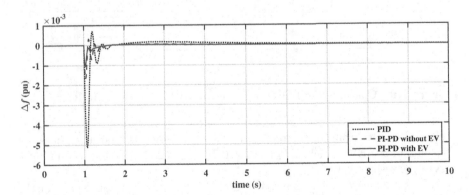

Figure 7.7   Comparative FDRs of the islanded MG considering 1% SLP

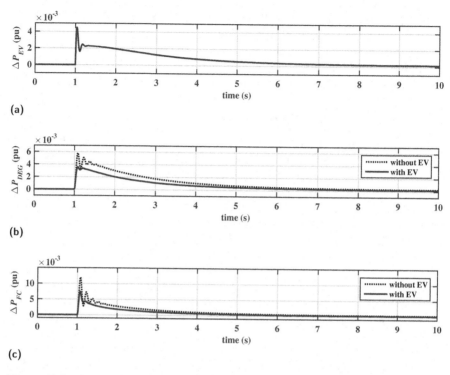

(a)

(b)

(c)

Figure 7.8 Power outputs (in pu) of: (a) the EV, (b) the DEG, and (c) the FC considering 1% SLP

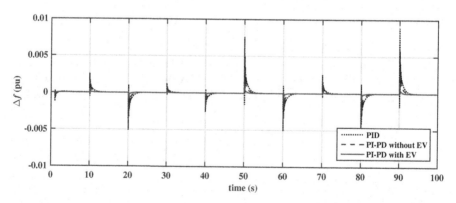

Figure 7.9 Comparative FDRs of the islanded MG considering RLP

self-explanatory from the figure. Figure 7.8 (a)–(c) show the variation in power outputs of the EV model (ΔP_{EV}), the DEG unit (ΔP_{DEG}), and the FC unit (ΔP_{EV}), respectively.

In another case, a random load perturbation (RLP) is considered in the MG while ΔP_{WTG} and ΔP_{PV} are kept constant this time also. Comparative FDRs of the islanded MG are shown in Fig. 7.9 with and without considering the EVs. Figure 7.10 (a) shows the RLP pattern in pu. Figs. 7.10 (b)-(d) show the variation in the ΔP_{EV}, ΔP_{DEG}, and ΔP_{EV}, respectively.

Further, the islanded MG is subjected to simultaneous multiple disturbances that include the RLP (as considered in the previous case), variation in the wind power and the solar power (i.e., RLP + ΔP_{WTG} + ΔP_{PV}). Comparative FDRs of the MG are shown in Fig. 7.11 with and without considering the EVs. Figure 7.12 (a) shows the multiple disturbances in pu. Figure 7.12 (b)–(d) depict the power outputs ΔP_{EV}, ΔP_{DEG}, and ΔP_{EV}, respectively, for this load disturbance condition.

Aggregating the results of all the load disturbance conditions considered in the islanded MG, it can be concluded that the PI-PD cascade controller performs better compared with conventional PID controller. Oscillations in the

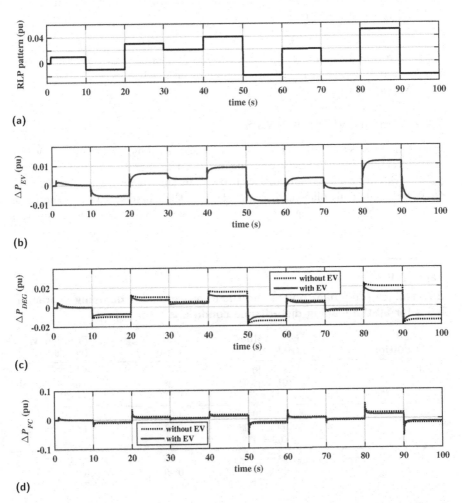

Figure 7.10 Power outputs (in pu) of: (a) the EV, (b) the DEG, and (c) the FC considering RLP

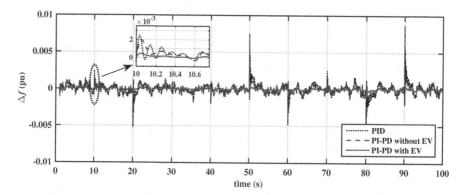

Figure 7.11 Comparative FDRs of the islanded MG considering simultaneous multiple disturbances

FDRs are rapidly damped out with the implementation of the cascade controller, and hence, the LFC requirements are satisfied. Also, when considering the EVs, the dynamic responses of the MG are appreciably improved.

7.4.2 SENSITIVITY ANALYSIS

For any control scheme, it is necessary to check its robustness when the system parameters are varied. This is obvious because the operating conditions in the power system keep on changing with time. In this regard, robustness of the PI-PD cascade controller against variation in the islanded MG parameters is tested here by performing the sensitivity analysis. For this, the parameters

Table 7.1

Eigenvalues (dominant poles) and their corresponding damping characteristics for all the loading disturbance conditions

Loading condition	Eigenvalues (poles)	Damping ratio (τ)	Frequency (rad/s)
1% SLP	$-1.88 + j0$	1.00	1.88
	$-5.61 \pm j0.94$	0.98	5.68
	$-21.23 \pm j52.71$	0.37	56.82
RLP	$-1.27 + j0$	1.00	1.27
	$-3.99 \pm j0.98$	0.97	4.11
	$-14.62 \pm j48.14$	0.29	50.33
RLP + ΔP_{WTG} + ΔP_{PV}	$-1.83 + j0$	1.00	1.83
	$-6.05 \pm j0.52$	0.99	6.07
	$-32.29 \pm j73.08$	0.41	79.87

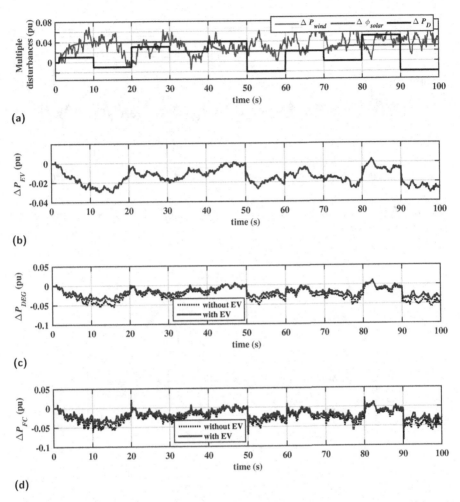

Figure 7.12 Power outputs (in pu) of: (a) the EV, (b) the DEG, and (c) the FC considering simultaneous multiple disturbances

D, H, and R are varied by $+$ 50% from their nominal value. The resulting FDR of the MG is plotted against that obtained when considering nominal values of these parameters. This is shown in Fig. 7.13. A careful observation of the FDRs reveals that they are more or less the same, hence justifying a robust behavior of the cascade controller. Further, this also justifies that the parameters of the cascade controller need not to be changed when subjected to variable operating conditions in the MG, hence proving it to be robust.

7.4.3 STABILITY ANALYSIS

The stability of the islanded MG for all the above-presented load disturbance conditions is also assessed via the root locus plots of the dominant poles. The

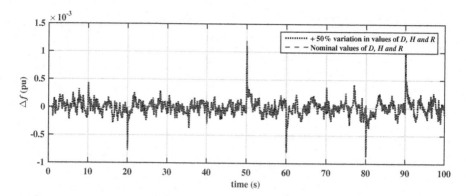

Figure 7.13 Comparative FDRs of the islanded MG corresponding to the sensitivity analysis

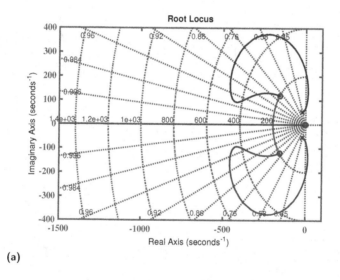

(a)

dominant poles obtained for each of the operating scenarios are presented in Table 7.1 along with their damping characteristics. A negative value of the real part in all the dominant poles justify a stable operation of the islanded MG under all the load disturbance conditions. Furthermore, to strengthen the stability analysis, root locus plots corresponding to all the loading conditions are shown in Fig. 7.14.

7.5 SUMMARY

This chapter discusses the implementation of an SSA-optimized PI-PD cascade controller for the LFC analysis of an islanded MG. Initially transfer

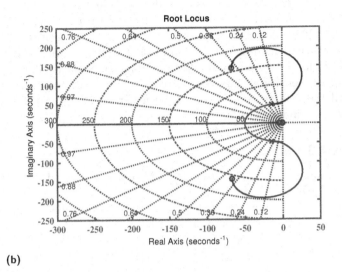

(b)

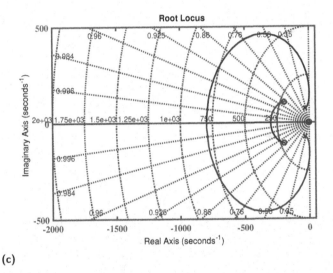

(c)

Figure 7.14 Root locus plots of the dominant poles corresponding to: a) 1% SLP, b) RLP, and c) simultaneous multiple disturbances

function modeling of the islanded MG and its various subsystems is minutely discussed. Emphasis is given on how the integration of the EVs with the MG can help in improving the LFC performance (dynamic response) of the islanded MG. Modeling of the PI-PD cascade controller and optimization of its tuning parameters through the SSA is also discussed. In the end, the competence of the cascade controller is demonstrated over the conventional PID controller considering three different load disturbance scenarios in the islanded MG. It has also been demonstrated that how integration of the EVs helps in improving the dynamic responses of the MG. Further, robustness of the controller has been validated against parametric variations in the MG by performing the sensitivity analysis. Lastly, stability analysis of the islanded MG for all the loading conditions has been performed using the root locus plots of the dominant poles.

QUESTIONS

1. In the dynamic model of the islanded MG shown in Fig. 7.1 try integrating different ESUs and compare the FDRs.
2. To optimize the parameters of the PI-PD cascade controller try implementing any other MOA and compare the FDRs.
3. For the aggregate EV model considered in the islanded MG try changing the values of $N_{control}(t)$, $N_{initial}$, $N_{control-in}(t)$, and $N_{plug-out}(t)$ and compare the FDRs.

8 LFC Study of a Grid-Connected Microgrid

The following chapter examines the LFC performance of a grid-connected microgrid (GcMG). Dynamic modeling of the GcMG is briefed initially followed by mathematical modeling of an FOPID plus double derivative (FOPID+DD) controller implemented for the LFC. Steps for optimization of the controller parameters by using the sine cosine algorithm (SCA) are outlined further. Later, time-domain simulation results are discussed demonstrating the potency of the controller. Finally, the robustness of the controller is validated against parametric variations and stability analysis of the GcMG is also performed via the Bode plots.

8.1 INTRODUCTION

In the grid-connected mode, an MG is engaged solely in exchanging P and Q with the utility grid. The V and f controls are taken care of by the utility grid itself owing to the high inertia of the conventional generators. In order to satisfy the gradually increasing load demand of the community it is advantageous to operate an MG in coordination with the utility grid so that dependence on the fossil fuel consumption can be mitigated. Whenever any type of disturbance condition persists in either the utility grid or the MG, the static switch at the PCC is opened and consequently the MG is disconnected from the utility grid and operated in the islanded mode. This chapter deals with the LFC study of a GcMG where diverse DG sources and ESUs operate in synchronism with a conventional reheat thermal power plant (RTPP).

Dynamic model of the GcMG is presented followed by the mathematical modeling of the SCA-optimized FOPID+DD controller. Finally, the efficacy of the controller and its robustness is justified through time-domain simulation results along with the stability analysis of the GcMG.

8.2 DYNAMIC MODEL OF THE GCMG

Figure 8.1 shows the dynamic model of the investigated GcMG [227]. The model consists of several DG sources namely WTG, DEG, FC, and AE and a BESU as the ESU that operate in synchronism with an RTPP. The RTPP combines a governor, a turbine, and a reheater in its model. The AE consumes some fraction of the WTG power output to generate hydrogen which is further utilized as an input to the FC unit to produce electrical power as output. The models of various PECs are ignored here in order to simplify the LFC analysis. The FOPID+DD controller is implemented as the secondary controller to

DOI: 10.1201/9781003477136-8

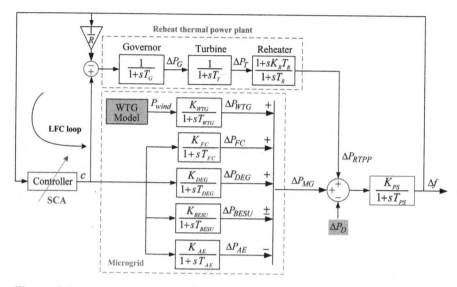

Figure 8.1 Dynamic model of the GcMG

regulate the power output of all the power generating units except the RES-based WTG unit because of the reasons outlined in Chapter 7.

All the units are represented by first-order transfer function models for the same reason as highlighted in the previous chapter as [227, 228]:

$$G_{WTG}(s) = \frac{\Delta P_{WTG}}{P_{wind}} = \frac{K_{WTG}}{1 + sT_{WTG}} \tag{8.1}$$

$$G_{FC}(s) = \frac{\Delta P_{FC}}{c} = \frac{K_{FC}}{1 + sT_{FC}} \tag{8.2}$$

$$G_{DEG}(s) = \frac{\Delta P_{DEG}}{c} = \frac{K_{DEG}}{1 + sT_{DEG}} \tag{8.3}$$

$$G_{BESU}(s) = \frac{\Delta P_{BESU}}{c} = \frac{K_{BESU}}{1 + sT_{BESU}} \tag{8.4}$$

$$G_{AE}(s) = \frac{\Delta P_{AE}}{c} = \frac{K_{AE}}{1 + sT_{AE}} \tag{8.5}$$

$$G_{FC}(s) = \frac{\Delta P_{FC}}{c} = \frac{K_{FC}}{1 + sT_{FC}} \tag{8.6}$$

$$G_{RTPP}(s) = \frac{\Delta P_{RTPP}}{c - \frac{\Delta f}{R}} = \left(\frac{1}{1 + sT_G}\right)\left(\frac{1}{1 + sT_T}\right)\left(\frac{1 + sK_R T_R}{1 + sT_R}\right) \tag{8.7}$$

Parameter values of the GcMG are given in **Appendix A**.

8.3 CONTROLLER MODEL AND OPTIMIZATION OF ITS PARAMETERS

This section examines the transfer function model of the FOPID+DD controller that is utilized as a secondary controller in the investigated GcMG. Further, parameters of the controller are optimized by implementing the SCA. Consequently, the steps involved in the optimization process are also portrayed.

8.3.1 MODELING OF THE CONTROLLER

It is well established in the quality literature that the controllers with a double derivative term exhibit better performance as compared to the controllers with only a single derivative term [227, 229]. In this regard, a conventional FOPID controller has been equipped with an additional double derivative (DD) term and hence the name FOPID+DD has been derived. Basics of the FO calculus have already been discussed in Chapter 6 of this book. The conventional FOPID controller is based on the FO calculus. Additionally, a double derivative term has been embedded in it. The proposed FOPID+DD controller is utilized as a secondary controller in the investigated GcMG to obtain its FDRs. The controller regulates the power output of the conventional RTPP and various DG sources and ESUs that include DEG, FC, AE, and BESU of the MG as per the input signal received (Δf). Transfer function model of the FOPID+DD controller is shown in Fig. 8.2. Here, K_p, K_i, K_d, and K_{dd} are the proportional, integral, derivative, and double derivative gains of the controller, respectively. λ and μ are the orders of the integrator and the differentiator, respectively. Output (c) of the controller, with input ($u = \Delta f$), is given by

$$c(t) = \left(K_p + K_i Q_t^{-\lambda} + K_d Q_t^{\mu} + K_{dd} Q_t^2\right) u(t) \tag{8.8}$$

Taking the Laplace transform of the above equation gives the transfer function of the controller as

$$G_{FOPID+DD}(s) = \frac{C(s)}{U(s)} = K_p + K_i s^{-\lambda} + K_d s^{\mu} + K_{dd} s^2 \tag{8.9}$$

In order to design the FOPID+DD controller, the *FOMCON* toolbox readily available in MATLAB/Simulink has been utilized [230]. The toolbox provides a very user-friendly interface to simulate numerous fractional order controllers for diverse applications. *FOMCON* stands for *fractional order modeling and control*. Another toolbox that can be readily implemented is the *Ninteger* toolbox [231].

8.3.2 OPTIMIZATION OF THE CONTROLLER PARAMETERS

In order to optimize the parameters of the FOPID+DD controller namely K_p, K_i, K_d, K_{dd}, λ, and μ, the SCA, as the MOA, has been implemented. The

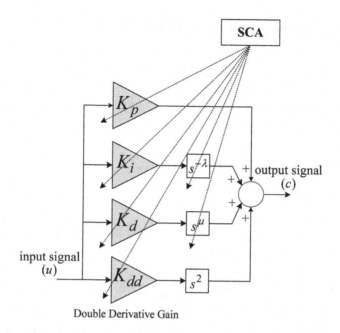

Figure 8.2 Transfer function model of the FOPID + DD controller

SCA was proposed by Seyedali Mirjalili in the year 2016 [232]. The algorithm is simple yet powerful population-based MOA for solving optimization problems. The algorithm operates utilizing a mathematical model based on sine and cosine functions. Similar to all the other MOAs, the SCA also runs by initializing a set of potential solutions in a *dim*-dimensional search space. These potential solutions are then guided to explore and exploit the search space by utilizing the sine-cosine dependent mathematical model and certain adaptive and random variables. For further details about the SCA, the interested readers may refer to [232].

An *ITSE*-based objective function (OF_{ITSE}) is utilized here in order to optimize the parameters of the FOPID+DD controller. The *ITSE* exhibits advantages of both the *ISE* and *ITAE* indices as multiplication of both time and squared error is involved. The LFC problem is thus formulated as Minimize:

$$OF_{ITSE} = \int_0^{t_{sim}} t\left[\Delta f^2\right] dt \qquad (8.10)$$

Subject to:

$$\begin{cases} K_{pmin} \leq K_p \leq K_{pmax} \\ K_{imin} \leq K_i \leq K_{imax} \\ K_{dmin} \leq K_d \leq K_{dmax} \\ K_{ddmin} \leq K_{dd} \leq K_{ddmax} \\ \lambda_{min} \leq \lambda \leq \lambda_{max} \\ \mu_{min} \leq \mu \leq \mu_{max} \end{cases} \tag{8.11}$$

where $K_{(p,i,d,dd)min}$ and $K_{(p,i,d,dd)max}$ are the lb and ub of the FOPID+DD controller parameters, respectively. λ_{min} and λ_{max} are the lb and ub, respectively, of the order of the integrator whereas μ_{min} and μ_{max} are the lb and ub, respectively of the order of the differentiator of the controller. After performing several trial runs, the range for the parameters K_p, K_i, K_d, and K_{dd} is ascertained in $[0, 10]$, whereas for λ and μ, the range is found in $[0, 2]$. t_{sim} denotes the simulation time in seconds (s). Below are the steps to optimize the FOPID+DD controller parameters using the SCA.

% Optimization of the FOPID+DD controller parameters using the SCA

% specifying the inputs

- $popsize$ = 40; dim = 6; t_{max} = 100; λ_{min} = 0; λ_{max} = 2;
- $\hat{a}$ = constant; μ_{min} = 0; μ_{max} = 2; $K_{min(p,i,d,dd)}$ = 0; $K_{max(p,i,d,dd)}$ = 10;

% initializing a population (set) of potential solutions

- $K_p = K_{min(p)} + rand(popsize, 1) \times (K_{max(p)} - K_{min(p)})$;
- $K_i = K_{min(i)} + rand(popsize, 1) \times (K_{max(i)} - K_{min(i)})$;
- $K_d = K_{min(d)} + rand(popsize, 1) \times (K_{max(d)} - K_{min(d)})$;
- $K_{dd} = K_{min(dd)} + rand(popsize, 1) \times (K_{max(dd)} - K_{min(dd)})$;
- $\lambda = \lambda_{min} + rand(popsize, 1) \times (\lambda_{max} - \lambda_{min})$;
- $\mu = \mu_{min} + rand(popsize, 1) \times (\mu_{max} - \mu_{min})$;

% evaluating the objective function value

- evaluate the objective function (OF_{ITSE}) corresponding to each initialized set of search agents using (8.10)

```
% start of iterations
for t = 1:t_max
```
- update the adaptive variable r_1 using
  $r_1 = \widehat{a} - \widehat{a} \times \frac{t}{t_{max}}$
- update the random variables r_2, r_3, and r_4 using
  $r_2 = 2\pi \times rand()$
  $r_3 = 2 \times rand()$
  $r_4 = rand()$
- update the position of each of the search agents using

$$h_j(t+1) = \begin{cases} h_j(t) + r_1 \times sin(r_2) \times |r_3 Q_j(t) - h_j(t)|, & if \ r_4 < 0.5 \\ h_j(t) + r_1 \times cos(r_2) \times |r_3 Q_j(t) - h_j(t)|, & if \ r_4 \geq 0.5 \end{cases}$$

```
end
```

```
% check for violation of boundaries
for i = 1:popsize
```

- if $h_j^i(t+1) > ub_j$
  then $h_j^i(t+1) = ub_j$

- elseif $h_j^i(t+1) < ub_j$
  then $h_j^i(t+1) = lb_j$
  else

```
end
```

```
% termination criterion
```
- repeat all the steps till t reaches t_{max}
- print the search agent with minimum value of the OF_{ITSE} and corresponding values of the K_p, K_i, K_d, K_{dd}, λ, and μ

Here, $h_j(t)$ is the position of the search agent at the tth iteration in the jth dimension, Q_j is the position of the destination point in the jth dimension, $| \ldots |$ denotes the absolute value, ub_j and lb_j are the upper and lower bounds of the jth dimension, respectively. The adaptive variable r_1 assists in balancing between exploration and exploitation of the search space, the random variable r_2 controls the magnitude of movement of the search agent towards or away from the Q, and the random variable r_3 controls the position of the Q by assigning it a random weight. The random variable r_4 helps in fairly switching between the sine and cosine functions. Key features of the SCA can be summarized as below [182]:

- Compared to the individual-based algorithms, the SCA generates and utilizes a set of potential solutions (search agents), thus enhancing its exploration capability and mitigating its chances of being getting stuck in a local optimum.
- Probable potential regions of the search space can be comfortably exploited resulting from different output values of the sine and cosine functions between -1 and 1.
- A destination point as the best approximation of the global optimum solution is stored in each of the iterations and is never discarded during the course of the iterations.
- The adaptive variable r_1 assists the SCA in uniformly transiting between exploration and exploitation of the search space.

8.4 SIMULATION RESULTS

Two different load disturbance conditions in the GcMG are considered here. One is considering multiple disturbances simultaneously, i.e., variations in ΔP_D, ΔP_{WTG}, and ΔP_{PV}. The other load disturbance condition considers these multiple disturbances along with taking into account the non-linearities associated with the RTPP that include generation rate constraint (GRC), governor dead band (GDB), and time delay (TD). FDRs of the GcMG are obtained corresponding to each of the load disturbance conditions with the FOPID+DD controller and the FOPID and the PID+DD controllers. Stability of the GcMG implemented with the FOPID+DD controller for both the load disturbance conditions is assessed in terms of the Bode plots. Robustness of the FOPID+DD controller against the parametric variations in the GcMG is further validated by performing the sensitivity analysis. A simulation time of 90 s is considered. SCA optimized parameters of the FOPID+DD controller correspond to $K_p = 5.24$, $K_i = 2.85$, $K_d = 8.11$, $K_{dd} = 1.21$, $\lambda = 1.47$, and $\mu = 0.96$. These parameters have been kept the same while considering both the operating conditions in the GcMG as well as for obtaining the Bode plots and for performing the sensitivity analysis.

8.4.1 COMPARATIVE FDRS OF THE GcMG

Figure 8.3 shows the comparative FDRs of the GcMG when considering the multiple disturbances (i.e., $\Delta P_D + \Delta P_{WTG} + \Delta P_{PV}$) in it. The FDRs shown are obtained with the FOPID+DD, FOPID, and PID+DD controllers. Figure 8.4 shows the power sharing between the RTPP and the MG subject to the above considered multiple disturbances. It can be clearly observed from the figure that the DG sources are utilized to their maximum when feeding the load demand as is done in case of a GcMG.

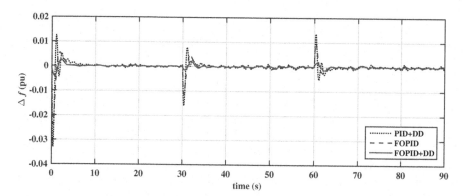

Figure 8.3 Comparative FDRs of the GcMG considering multiple disturbances only

8.4.2 COMPARATIVE FDRS OF THE GcMG CONSIDERING RTPP NON-LINEARITIES

Figure 8.5 shows the dynamic model of the GcMG when non-linearities related to the RTPP are considered. The confronted non-linearities are GRC of 5%, GDB of 0.0006 pu and TD of 200 ms. The comparative FDRs of the GcMG are shown in Fig. 8.6. Power sharing between the RTPP and the MG is shown in Fig. 8.7 where also the DG sources are utilized to their maximum.

8.4.3 SENSITIVITY ANALYSIS

To validate the robustness of the FOPID+DD controller against the variations in the various parameters of the GcMG a sensitivity analysis is performed. Here variation in certain parameters is considered that include +30% in T_G,

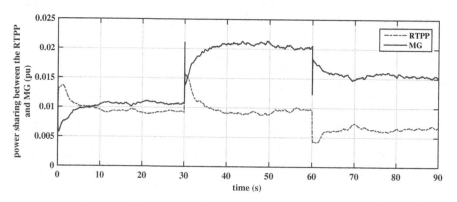

Figure 8.4 Power sharing between the RTPP and MG when considering multiple disturbances only in the GcMG

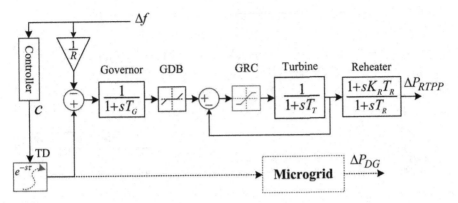

Figure 8.5 Dynamic model of the GcMG considering non-linearities in the RTPP

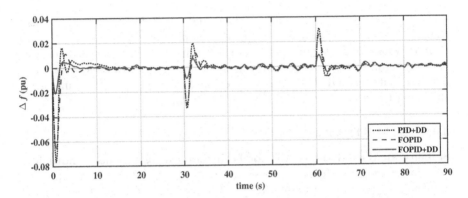

Figure 8.6 Comparative FDRs of the GcMG considering multiple disturbances and the RTPP non-linearities

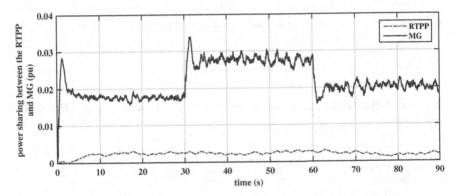

Figure 8.7 Power sharing between the RTPP and MG when considering multiple disturbances and non-linearities in the GcMG

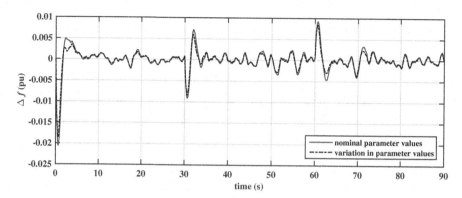

Figure 8.8 Comparative FDRs of the GcMG considering variation in different parameters and their nominal values

-25% in T_T, +40% in T_{FC}, -30% in T_{AE}, and +25% in R. The comparative FDRs of the GcMG taking into account these parametric variations and with nominal values of these parameters are shown in Fig. 8.8. It can be concluded from the figure that both the FDRs resemble each other to a great extent that justifies that the FOPID+DD controller operates in a robust manner when subjected to parametric variations in the system.

8.4.4 STABILITY ANALYSIS

Stability analysis of the GcMG for both the load disturbance conditions is done by using the Bode plots. These are shown in Fig. 8.9. The corresponding gain margin (GM) in decibels (dB), phase margin (PM) in degrees (deg), gain crossover frequency (ω_{gc}) in radians/sec (rad/s), and phase crossover frequency (ω_{pc}) in radian/sec (rad/s) are shown in Table 8.1. Since the GM and PM for both the load disturbance conditions of the GcMG are positive and also since PM > GM for both the conditions, this justifies a stable operation of the GcMG. Notable here is that in order to obtain the Bode diagrams, the

Table 8.1

Details of the Bode plots corresponding to the different load disturbance conditions

Loading condition	GM (dB)	ω_{pc}(rad/s)	PM (deg)	ω_{gc}(rad/s)
Multiple disturbances	8.98	6.26	36.3	4.97
Multiple disturbance + RTPP non-linearities	5.86	5.89	21.1	4.88

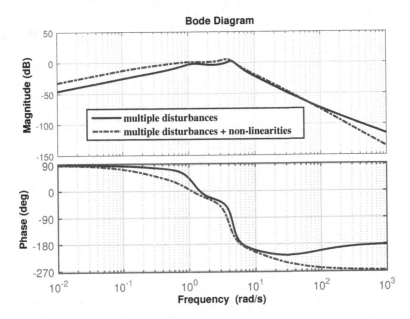

Figure 8.9   Bode plots corresponding to different load disturbance conditions in the GcMG

Simulink model of the GcMG was linearized and then all non-minimal state dynamics were eliminated.

8.5   SUMMARY

This chapter entails a comprehensive discussion on the LFC performance of a GcMG. The entire investigated system comprises certain DG and ES units connected to a conventional RTPP. FDRs of the GcMG are obtained considering two different load disturbance conditions and utilizing an FOPID+DD controller as the secondary controller. Further, robustness of the FOPID+DD controller is authenticated against variations in certain parameters of the GcMG by conducting the sensitivity analysis. Finally, the stability of the GcMG is evaluated for both the considered operating conditions using the Bode plot analysis.

QUESTIONS

1. In the dynamic model of the GcMG shown in Fig. 8.1 try integrating different DG sources and ESUs and compare the FDRs.
2. To optimize the parameters of the FOPID+DD cascade controller try implementing any other MOA and compare the FDRs.

3. Try to conduct the stability analysis of the GcMG by using the root locus method.
4. Try changing the values of the RTPP non-linearities, i.e., GDB, GRC, and TD and compare the FDRs.

9 LFC Study of a Multi-Microgrid System

The following chapter examines the LFC performance of a multi-microgrid (MMG) system. LFC study of such a system is obviously more demanding owing to the interconnection of two or more MGs whereby each comprising the highly fluctuating RES-based DG units. The configuration or structure of a general MMG system is initially dealt with followed by the detailed dynamic response modeling of a two-MG interconnected MMG system. A 2DoF-PID controller is implemented as a secondary controller in each of the MGs to obtain the FDRs. Mathematical modeling of the 2DoF-PID controller is further investigated followed by the steps for optimization of its parameters. The crow search algorithm (CSA) is utilized in this regard. Finally, the robustness of the controller is authenticated subject to parametric variations in the MMG system by performing the sensitivity analysis.

9.1 INTRODUCTION

An MMG system consists of several individual MGs interconnected to each other and to the utility grid via the PCC. Like the MG, an MMG system can also be operated either in islanded mode or grid-connected mode. An MMG system possesses the following advantages [233]:

1. *Improved resilience:* MMG systems are intrinsically more resilient than single MG systems or conventional utility grids. If one MG experiences a problem, the others are able to continue operating independently. This redundancy reduces the likelihood of total power interruptions during disruptions such as natural disasters and equipment failures.

2. *Non-dependency on the utility grid:* MGs within an MMG system can function independently of the utility grid, decreasing reliance on centralized power generation. This autonomy can provide energy security in the event of grid outages or shortages.

3. *A step toward smart grid:* MMG systems often incorporate sophisticated monitoring, control, and communication technologies, thereby becoming part of the expanding smart grid framework. This allows for data analysis and grid management on real-time basis.

Table 9.1

Details of the MMG system setups worldwide

Name	Country	AC/DC	Single/three ph	Voltage level	Application status
CERTS	USA	AC	three ph	LV	experimental setup
NREL	USA	AC/DC	three ph	LV	experimental setup
IIT	USA	DC	–	LV	experimental setup
Sendai	Japan	AC/DC	three ph	MV	experimental setup
Shimizu	Japan	AC/DC	three ph	MV	experimental setup
CESI	Italy	AC	three ph	LV	experimental setup
Labein	Spain	AC/DC	single ph	LV	experimental setup
ARMINES	France	AC	single/three ph	LV	experimental setup
Luxi Island	China	AC	three ph	MV	remote area
Wanshan Islands	China	AC	three ph	MV	remote area
Kythnos	Greece	AC	three/single ph	LV	remote area
MVV	Germany	AC	three ph	LV	experimental setup
Demotec	Germany	AC	three/single ph	LV	experimental setup

4. *Enhanced reliability:* An MMG system can strategically position DG sources to improve grid reliability. This is especially essential in remote or critical infrastructure locations where grid stability is crucial.

5. *Better stability and controllability:* By utilizing a hierarchical control structure, the stability and controllability of an MMG system can be significantly enhanced. This hierarchical control approach involves organizing and managing the various components and levels within the MMG system in a structured manner.

However, compared to the grid-connected single MGs, implementation of the control and protection schemes is more complex. Further, for the above-stated advantages to be fully realized, effective planning, coordination, and supervision are required.

Regarding the application status of the MMG system around the world, it is possible to declare here that the countries in Europe, the United States, Japan, and other developed countries have reached a mature stage in this respect. On the other hand, the idea is still in its formative stages in nations such as China, which are still considered to be developing countries. The MMG setups from around the world are detailed in Table 9.1. [234].

Figure 9.1 shows the structure of a general MMG system. The system comprises of n identical MGs connected to the utility grid via the PCC. Each MG consists of some DG sources and ESUs connected to the AC bus through suitable PECs.

9.2 DYNAMIC MODEL OF THE MMG SYSTEM

Figure 9.2 shows the dynamic model of a two-MG interconnected MMG system [235]. Additionally, each MG has an aggregate EV model integrated to analyse how it affects the LFC performance of the MMG system. Models of the various PECs are disregarded in order to make the frequency dynamics of

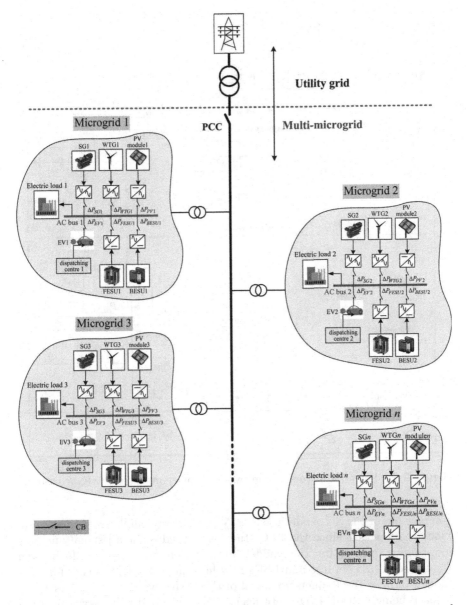

Figure 9.1 Structure of a general MMG system comprising n interconnected MGs

the proposed MMG easier to grasp. Thus, the PV, WTG, BESU, and FESU linearized first-order transfer function models make up the dynamic model. While bigger capacity MGs use turbine-driven generators, lesser capacity MGs often use diesel engine generators as SG. The governor and turbine are two

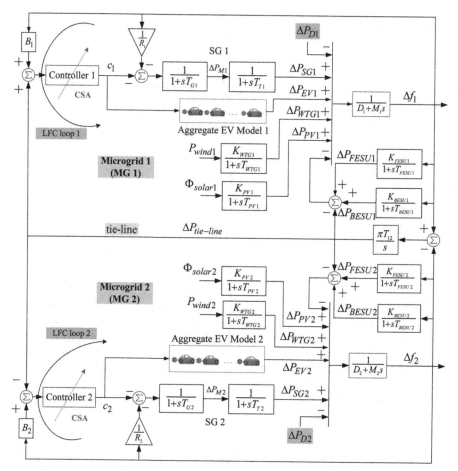

Figure 9.2 Dynamic model of the two-MG interconnected MMG system

components of the SG model used [235]. As already emphasized in the previous chapter that since the LFC study is focused on small signal analysis, lower-order transfer function models are sufficient to represent them, even though some DG sources and ESUs may have higher-order transfer function models [6]. For the frequency control of the MMG system, controllers 1 and 2 (each being a 2DoF-PID controller), that correspond to the two interconnected MGs, control the power output of the SG unit and the aggregate EV model in each MG. By exchanging power within the MMG, ESUs like the BESU and the FESU function as local active power sources to stabilize frequency variations. Linearized first-order transfer function models of different DG sources and ESUs are given as [8]:

$$G_{WTGi}(s) = \frac{\Delta P_{WTGi}}{P_{windi}} = \frac{K_{WTGi}}{1 + sT_{WTGi}} \qquad (9.1)$$

$$G_{PVi}(s) = \frac{\Delta P_{PVi}}{\Delta \Phi_{solari}} = \frac{K_{PVi}}{1 + sT_{PVi}} \qquad (9.2)$$

$$G_{SGi}(s) = \frac{\Delta P_{SGi}}{u_i} = \left(\frac{1}{1 + sT_{Gi}}\right)\left(\frac{1}{1 + sT_{Ti}}\right) \qquad (9.3)$$

$$G_{BESUi}(s) = \frac{\Delta P_{BESUi}}{\Delta f_i} = \frac{K_{BESUi}}{1 + sT_{BESUi}} \qquad (9.4)$$

$$G_{FESUi}(s) = \frac{\Delta P_{FESUi}}{\Delta f_i} = \frac{K_{FESUi}}{1 + sT_{FESUi}} \qquad (9.5)$$

for $i = 1$ and 2. For more details about the MMG system, interested readers may refer to [235]. Parameter values of the system are given in **Appendix A.**

9.3 CONTROLLER MODEL AND OPTIMIZATION OF ITS PARAMETERS

This section discusses a detailed mathematical modeling of the 2DoF-PID controller. Further, a detailed procedure for optimizing the parameters of the controller using the CSA is presented and hence the LFC problem is formulated.

9.3.1 MODELING OF THE CONTROLLER

The general scheme of a 2DoF control approach for a feedback control system has already been dealt with in Chapter 6 of this book. Compared to the conventional 1DoF control approach, a 2DoF control approach enhances the set-point tracking and load disturbance rejection capability. Mathematical model of a 2DoF-PID controller is shown in Fig. 9.3. Here, K_p, K_i, and K_d denote the proportional, integral, and derivative parameters of the controller. a and b correspond to the proportional and derivative set-point weights and N is the derivative filter coefficient. r, y, and c are the reference signal, feedback signal, and output/control signal, respectively. Laplace transform of the control signal, i.e., $U(s)$ is given by:

$$U(s) = K_p\left(aR(s) - Y(s)\right) + \frac{K_i}{s}\left(R(s) - Y(s)\right) + K_d s \frac{(bR(s) - Y(s))}{1 + \frac{K_d s}{K_p N}} \qquad (9.6)$$

where, $R(s)$ and $Y(s)$ are the Laplace transform of r and y, respectively.

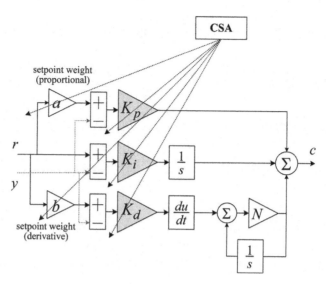

Figure 9.3 Mathematical model of the 2DoF-PID controller

9.3.2 OPTIMIZATION OF THE CONTROLLER PARAMETERS

In order to optimize the parameters of the 2DoF-PID controller, namely, K_p, K_i, K_d, a, b, and N, the CSA has been implemented. CSA a newly-developed population-based MOA inspired by the intelligent behavior of the crows developed by Askarzadeh [236]. The idea that crows hide their extra food in certain locations and collect it when necessary served as inspiration for the CSA. Since they have demonstrated some astounding intellect, crows are among the most intelligent birds in the world. Readers who are interested in learning more information about the CSA may refer [236].

In order to optimize the controller parameters, the OF_{ITSE} is utilized owing to its advantages as discussed in the previous chapter. The LFC problem is thus formulated as

Minimize:

$$OF_{ITSE} = \int_0^{t_{sim}} t \left[\Delta f_1^2 + \Delta f_2^2 + \Delta P_{tie-line}^2 \right] dt \tag{9.7}$$

Subject to:

$$\begin{cases} K_{pmin} \leq K_t \leq K_{pmax} \\ K_{imin} \leq K_i \leq K_{imax} \\ K_{dmin} \leq K_d \leq K_{dmax} \\ a_{min} \leq a \leq a_{max} \\ b_{min} \leq b \leq b_{max} \\ N_{min} \leq N \leq N_{max} \end{cases} \tag{9.8}$$

where $K_{(p,i,d)min}$ and $K_{(p,i,d)max}$ are the *lb* and *ub* of the 2DoF-PID controller parameters, respectively. a_{min} and a_{max}, b_{min} and b_{max}, and N_{min}

and N_{max} are the *lb* and *ub* of the proportional set-point, derivative set-point, and derivative filter coefficient of the controller, respectively. After performing several trial runs, the range for the parameters K_p, K_i, K_d, and K_{dd} is ascertained in $[0, 5]$, whereas for a, b, and N the range is set in $[0, 1]$ and $[0, 120]$, respectively. t_{sim} denotes the simulation time (s). Below are the steps to optimize the cascade controller parameters using the CSA.

% Optimization of the 2DoF-PID controller parameters using the CSA

% specifying the inputs

- $popsize$ = 20; dim = 6; t_{max} = 100; a_{min} = 0; a_{max} = 1; b_{min} = 0; b_{max} = 1;

- $K_{min(p,i,d)}$ = 0; $K_{max(p,i,d)}$ = 5; $flen$ = 2; $aprob$ = 0.1

% initializing a population of potential solutions (flock of $popsize$ crows)

- $K_p = K_{min(p)} + rand(popsize, 1) \times (K_{max(p)} - K_{min(p)})$;
- $K_i = K_{min(i)} + rand(popsize, 1) \times (K_{max(i)} - K_{min(i)})$;
- $K_d = K_{min(d)} + rand(popsize, 1) \times (K_{max(d)} - K_{min(d)})$;
- $a = a_{min} + rand(popsize, 1) \times (a_{max} - a_{min})$;
- $b = b_{min} + rand(popsize, 1) \times (b_{max} - b_{min})$;
- $N = N_{min} + rand(popsize, 1) \times (N_{max} - N_{min})$;

% evaluating the objective function value

- evaluate the objective function (OF_{ITSE}) corresponding to each initialized set of search agents (crows) using (9.7)

% start of iterations
for t = 1:t_{max}

for i = 1:$popsize$

- update the position of each of the search agents using

$$h_i(t+1) = \begin{cases} h_i(t) + r_i \times flen_i(t)) \times |Q_j(t) - h_i(t)|, \; if \; r_j \geq aprob_j(t) \\ (\text{State 1}) \\ a \; random \; number, \; otherwise \\ (\text{State 2}) \end{cases}$$

- evaluate the fitness of the new crow positions

- update the memory of each crow

end

% check for violation of boundaries
for i = 1:*popsize*

- if $\quad\quad h_i(t+1) > ub_i$
  then $\quad\quad h_i(t+1) = ub_i$

- elseif $\quad h_i(t+1) < ub_i$
  then $\quad\quad h_i(t+1) = lb_i$
  else

- end

end

% termination criterion
- repeat all the steps till t reaches t_{max}
- print the search agent with minimum value of the OF_{ITSE} and corresponding values of the K_p, K_i, K_d, a, b, and N

Here, $h_i(t)$ is the position of the ith crow at the tth iteration, Q_j is the hiding position of jth crow at the tth iteration, ub_i and lb_i are the upper and lower bounds of the ith dimension, respectively. r_i and r_j are uniform random numbers in the range $[0, 1]$. $flen_i(t)$ is the flight length of ith crow at the tth iteration and $aprob_j(t)$ is the awareness probability of jth crow at the tth iteration. It is noteworthy here that $flen$ and $aprob$ help in balancing between exploration and exploitation of the search space by the algorithm. A small value for these tends to exploit the search space more whereas a large value results in an enhanced exploration of the search space.

9.4 SIMULATION RESULTS

The competence of the CSA optimized 2DoF-PID controller is established here considering two different load disturbance conditions in the MMG system. Performance of the 2DoF-PID controller is compared with the conventional PID controller. Also, FDRs and tie-line power deviations of the MMG system excluding the aggregate EV model are obtained and compared to emphasize on the significance of the EVs. A simulation time of 150 s is considered. The CSA optimized parameters of the 2DoF-PID controller obtained are $K_p = 3.15$, $K_i = 4.32$, $K_d = 1.58$, $a = 0.59$, $b = 0.82$, and $N = 76.32$. These parameters have been kept the same while considering both the load disturbance conditions in the MMG system.

9.4.1 COMPARATIVE FDRs OF THE MMG SYSTEM

Figure 9.4(a)–(c) show the comparative FDRs of the MMG considering an RLP pattern in each MG simultaneously. The FDRs correspond to the frequency deviation in MG1, i.e., Δf_1, the frequency deviation in MG2, i.e., Δf_2, and the tie-line power deviation, i.e., $\Delta P_{tie-line}$. It can be clearly observed that the FDRs are improved with the 2DoF-PID controller as compared to the conventional PID controller. Further, the FDRs worsen when not considering the aggregate EV model.

Comparative FDRs of the MMG system considering a pulse load perturbation (PLP) pattern simultaneously in both the MGs are shown in Figs. 9.5(a)–(c). Competence of the 2DoF-PID controller is clearly revealed from the figures whereby Δf_1, Δf_2, and $\Delta P_{tie-line}$ are improved. Here also the FDRs without considering the EVs are aggravated, thus obviously italicizing the effect of their consideration. A PLP pattern with a period of 50 s and amplitude of 0.07 pu is considered for obtaining the FDRs.

Variation in the power output of the aggregate EV model (ΔP_{EV1} and ΔP_{EV2}) integrated in both the MGs of the MMG system is shown in Figs. 9.6(a)–(b). Figure 9.6(a) shows the power output variation for the RLP pattern while Fig. 9.6(b) shows it for the PLP pattern.

9.4.2 SENSITIVITY ANALYSIS

A sensitivity analysis is conducted to validate the robustness of the FOPID+DD controller against ±30% parametric variations of the MMG system. Variation in the following parameters is considered for both the interconnected MGs: +30% for T_{G1} and T_{G2}, -30% for T_{EV1} and T_{EV2}, +30% for T_{BESU1} and T_{BESU2}, and -30% for T_{FESU1} and T_{FESU2}. Figure 9.7 (a)–(c) depicts the comparative FDRs and tie-line power deviations of the MMG system that account for these parametric variations and have nominal values for these parameters considering the RLP pattern. From the figure, it can be deduced that all these FDRs closely resemble each other, which demonstrates that the 2DoF-PID controller is robust when subjected to parametric variations in the MMG system.

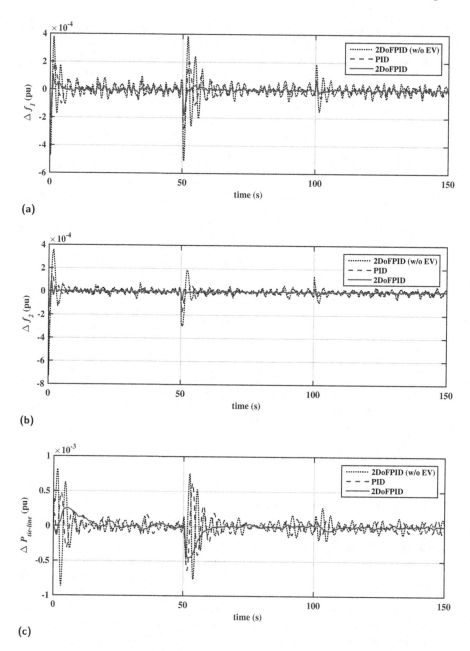

Figure 9.4 Comparative FDRs of the MMG system for the RLP pattern

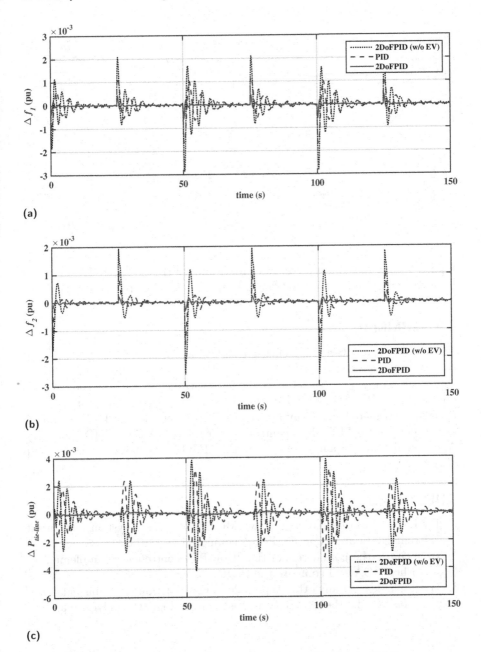

Figure 9.5 Comparative FDRs of the MMG system for the PLP pattern

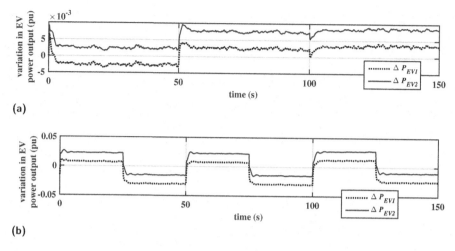

(a)

(b)

Figure 9.6 Variation in power outputs (in pu) of the aggregate EV model in both the MGs for (a) the RLP pattern and (b) the PLP pattern

9.5   SUMMARY

This chapter presents a comprehensive LFC analysis of an MMG system comprising of two interconnected MGs via a tie-line. A CSA-optimized 2DoF-PID controller is implemented as a secondary controller in each of the MGs to obtain the FDRs of the MMG system. Two different load disturbance conditions are considered, and comparative FDRs are obtained for both to exhibit the potency of the 2DoF-PID controller over the conventional PID controller. Additionally, integration of the EVs in the MMG is emphasized with the help of the FDRs obtained.

QUESTIONS

1. Try interconnecting more than two MGs in the MMG system integrating different DG sources and ESUs and compare the FDRs.
2. To optimize the parameters of the 2DoF-PID controller try implementing any other MOA and compare the FDRs.
3. Try implementing any other type of load perturbation (say sinusoidal load perturbation) in the MMG system shown in Fig. 9.2 and compare the FDRs.

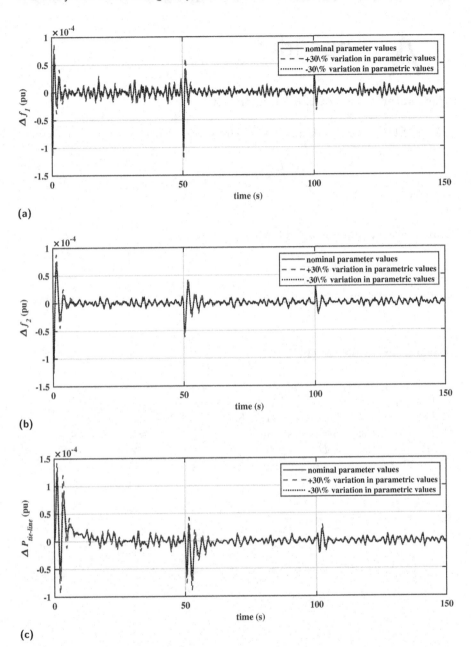

Figure 9.7 Comparative FDRs of the MMG system considering parametric variations and nominal parameter values for the RLP pattern

A Appendix

Parameter values of the islanded MG:

$2H = 0.1167$ s; $D = 0.015$; $T_T = 0.01$ s; $T_G = 0.4$ s; $T_{IN} = 0.04$ s; $T_{IC} = 0.004$ s; $T_{FC} = 0.28$ s; $T_{BESU} = 0.1$ s; $T_{FESU} = 0.1$ s; $R = 2.4$ pu Hz/MW; $T_{WTG} = 1.5$ s; $K_{WTG} = 1$; $T_{PV} = 1.8$ s; $K_{PV} = 1$; $T_{EV} = 1$ s; $C_{kW}^* = 3$; $C_{kWh}^* = 15$, $N_{control} = 3600$; $N_{initial} = 5000$; $N_{control-in} = 1100$; $N_{plug-out} = 2500$.

Parameter values of the GcMG:

$K_{PS} = 120$; $R = 2.4$ pu Hz/MW; $T_{PS} = 20$ s; $K_R = 0.5$; $T_T = 0.08$ s; $T_R = 10$ s; $T_G = 0.4$ s; $K_{WTG} = 1$; $K_{FC} = 0.01$; $T_{WTG} = 1.5$ s; $T_{FC} = 4$ s; $K_{DEG} = 0.003$; $K_{AE} = 0.002$; $T_{DEG} = 2$; $T_{AE} = 0.5$ s; $T_{BESU} = 0.1$ s; $K_{BESU} = -0.003$.

Parameter values of the MMG:

$K_{WTG1} = 1$; $K_{WTG2} = 1$; $T_{WTG1} = 0.5$ s; $T_{WTG1} = 0.5$ s; $K_{PV1} = 1$; $K_{PV2} = 1$; $T_{PV1} = 1.5$ s; $T_{PV2} = 1.5$ s; $K_{BESU1} = -3$; $K_{BESU2} = -4$; $T_{BESU1} = 0.1$ s; $T_{BESU2} = 0.1$ s; $K_{FESU1} = -1.5$; $K_{FESU2} = -2$; $T_{FESU1} = 0.1$; $T_{FESU1} = 0.1$; $T_{G1} = 0.1$ s; $T_{G2} = 0.1$ s; $T_{T1} = 0.4$ s; $T_{T2} = 0.4$ s; $T_{12} = 1.4$ s; $M1 = 8$ s; $M2 = 8$ s; $D1 = 1$; $D2 = 1$; $R1 = 0.05$ pu Hz/MW; $R2 = 0.04$ pu Hz/MW; $B1 = 10$; $B2 = 12.5$; $T_{EV1} = 1$ s; $T_{EV2} = 1$ s.

DOI: 10.1201/9781003477136-A

References

1. P. Kundur, J. Paserba, V. Ajjarapu, G. Andersson, A. Bose, C. Canizares, N. Hatziargyriou, D. Hill, A. Stankovic, C. Taylor *et al.*, "Definition and classification of power system stability ieee/cigre joint task force on stability terms and definitions," *IEEE transactions on Power Systems*, vol. 19, no. 3, pp. 1387–1401, 2004.
2. N. Hatziargyriou, J. Milanovic, C. Rahmann, V. Ajjarapu, C. Canizares, I. Erlich, D. Hill, I. Hiskens, I. Kamwa, B. Pal *et al.*, "Definition and classification of power system stability revisited & extended," *IEEE Transactions on Power Systems*, vol. 36, no. 4, pp. 3271–3281, 2020.
3. H. Bevrani, *Robust Power System Frequency Control*, 2nd ed., ser. Power Electronics and Power Systems. Springer International Publishing Switzerland, 2014.
4. P. M. Anderson, B. L. Agrawal, and J. E. Van Ness, *Subsynchronous resonance in power systems*. John Wiley & Sons, 1999, vol. 9.
5. H. Bevrani and T. Hiyama, *Intelligent Automatic Generation Control*. CRC Press, Taylor & Francis Group, 2011.
6. H. Bevrani, F. Habibi, P. Babahajyani, M. Watanabe, and Y. Mitani, "Intelligent frequency control in an AC microgrid: Online PSO-based fuzzy tuning approach," *IEEE Transactions on Smart Grid*, vol. 3, no. 4, pp. 1935–1944, 2012.
7. R. Ali, T. H. Mohamed, Y. S. Qudaih, and Y. Mitani, "A new load frequency control approach in an isolated small power systems using coefficient diagram method," *Electrical Power and Energy Systems*, vol. 56, pp. 110–116, 2014.
8. H. Bevrani, M. R. Feizi, and S. Ataee, "Robust frequency control in an islanded microgrid: H-∞ and μ-synthesis approaches," *IEEE Transactions on Smart Grid*, vol. 7, no. 2, pp. 706–717, 2016.
9. P. M. Anderson and M. Mirheydar, "A low-order system frequency response model," *IEEE Transactions on Power Systems*, vol. 5, no. 3, pp. 720–729, 1990.
10. K. Shimizu, T. Masuta, Y. Ota, and A. Yokoyama, "Load frequency control in power system using vehicle-to-grid system considering the customer convenience of electric vehicles," in *2010 International Conference on Power System Technology*, 2010, pp. 1–8.
11. Shimizu, Masuta, Ota, and Yokoyama, "A new load frequency control method in power system using vehicle-to-grid system considering users' convenience," in *Proceedings of the 17th Power System Computation Conference, Stockholm, Sweden*, 2011, pp. 22–26.
12. T. Masuta and A. Yokoyama, "Supplementary load frequency control by use of a number of both electric vehicles and heat pump water heaters," *IEEE Transactions on Smart Grid*, vol. 3, no. 3, pp. 1253–1262, 2012.
13. Pahasa and Ngamroo, "Coordinated control of wind turbine blade pitch angle and PHEVs using MPCs for load frequency control of microgrid," *IEEE Systems Journal*, vol. 10, no. 1, pp. 97–105, 2014.

14. M. H. Khooban, T. Niknama, F. Blaabjerg, and T. Dragicevic, "A new load frequency control strategy for micro-grids with considering electrical vehicles," *Electric Power Systems Research*, vol. 143, pp. 585–598, 2016.

15. M. H. Khooban, T. Niknam, M. Shasadeghi, T. Dragicevic, and F. Blaabjerg, "Load frequency control in microgrids based on a stochastic non-integer controller," *IEEE Transactions on Sustainable Energy*, Vol. 9, no. 2, pp. 853–861, 2017.

16. S. Debbarma and A. Dutta, "Utilizing electric vehicles for LFC in restructured power systems using fractional order controller," *IEEE Transactions on Smart Grid*, vol. 8, no. 6, pp. 2554–2564, 2017.

17. B. Khokhar, S. Dahiya, and K. P. S. Parmar, "A robust cascade controller for load frequency control of a standalone microgrid incorporating electric vehicles," *Electric Power Components and Systems*, vol. 48, no. 6-7, pp. 711–726, 2020.

18. T. Senjyu, T. Nakaji, K. Uezato, and T. Funabashi, "A hybrid power system using alternative energy facilities in isolated island," *IEEE Transactions on Energy Conversion*, vol. 20, no. 2, pp. 406–414, 2005.

19. D. J. Lee and L. Wang, "Small-signal stability analysis of an autonomous hybrid renewable energy power generation/energy storage system part I: Time-domain simulations," *IEEE Transactions on Energy Conversion*, vol. 20, no. 1, pp. 311–320, 2008.

20. M.-H. Khooban, "Secondary load frequency control of time-delay stand-alone microgrids with electric vehicles," *IEEE Transactions on Industrial Electronics*, vol. 65, no. 9, pp. 7416–7422, 2018.

21. M.-H. Khooban, T. Dragicevic, F. Blaabjerg, and M. Delimar, "Shipboard microgrids: A novel approach to load frequency control," *IEEE Transactions on Sustainable Energy*, vol. 9, no. 2, pp. 843–852, 2018.

22. B. Khokhar, S. Dahiya, and K. P. S. Parmar, "A novel fractional order proportional integral derivative plus second-order derivative controller for load frequency control," *International Journal of Sustainable Energy*, vol. 40, no. 3, pp. 235–252, 2020.

23. B. Khokhar, S. Dahiya, and K. S. Parmar, "A novel hybrid fuzzy PD-TID controller for load frequency control of a standalone microgrid," *Arabian Journal for Science and Engineering*, vol. 46, no. 2, pp. 1053–1065, 2021.

24. Khokhar, Dahiya, and Parmar, "Load frequency control of a microgrid employing a 2D sine logistic map based chaotic sine cosine algorithm," *Applied Soft Computing*, vol. 109, pp. 1–17, 2021.

25. I. Pan and S. Das, "Fractional order AGC for distributed energy resources using robust optimization," *IEEE Transactions on Smart Grid*, Vol. 7, no. 5, pp. 2175–2186, 2015.

26. Pan and Das, "Fractional order fuzzy control of hybrid power system with renewable generation using chaotic PSO," *ISA Transactions*, vol. 62, pp. 19–29, 2016.

27. J. Liu, Q. Yao, and Y. Hu, "Model predictive control for load frequency of hybrid power system with wind power and thermal power," *Energy*, vol. 172, pp. 555–565, 2019.

28. "Global energy-related carbon dioxide emissions by source, 1990-2018, IEA, paris," IEA, Tech. Rep., 2019. [Online].

Available: https://www.iea.org/data-and-statistics/charts/global-energy-related-carbon-dioxide-emissions-by-source-1990-2018

29. Paris, "Renewables 2021," IEA, Tech. Rep., 2021. [Online]. Available: https://www.iea.org/reports/renewables-2021

30. P. A. Owusu and S. Asumadu-Sarkodie, "A review of renewable energy sources, sustainability issues and climate change mitigation," *Cogent Engineering*, vol. 3, no. 1, p. 1167990, 2016.

31. K. Thoubboron, "Advantages and disadvantages of renewable energy," Nov. 2021. [Online]. Available: https://news.energysage.com/advantages-and-disadvantages-of-renewable-energy/

32. "Installed capacity report," Central Electricity Authority, Government of India, Tech. Rep., 2023.

33. IEA, "Global energy review 2019," IEA, Paris, Tech. Rep., 2019.

34. "World energy outlook 2021," International Energy Agency (IEA), Paris, Tech. Rep., 2021. [Online]. Available: https://www.iea.org/reports/world-energy-outlook-2021

35. "Renewable power generation by technology in the net zero scenario, 2010-2030," IEA, Paris, Tech. Rep., 2022.

36. IRENA, "Renewable capacity statistics 2023," International Renewable Energy Agency (IRENA), Abu Dhabi, Tech. Rep., 2023.

37. L. Fusheng, L. Ruisheng, and Z. Fengquan, *Microgrid Technology and Engineering Application*. China Electric Power Press, Elsevier Inc., 2016.

38. Office of Energy Efficiency & Renewable Energy, "Types of hydropower plants." [Online]. Available: https://www.energy.gov/eere/water/types-hydropower-plants

39. "IEA (2021), geothermal power," IEA, Paris, Tech. Rep., 2021.

40. M. Guo, W. Song, and J. Buhain, "Bioenergy and biofuels: History, status, and perspective," *Renewable and sustainable energy reviews*, vol. 42, pp. 712–725, 2015.

41. "IRENA (2020), innovation outlook: Ocean energy technologies," International Renewable Energy Agency, Abu Dhabi, Tech. Rep., 2020.

42. IRENA, "Renewable capacity statistics 2022," International Renewable Energy Agency (IRENA), Abu Dhabi, Tech. Rep., 2022.

43. S. R. Sinsel, R. L. Riemke, and V. H. Hoffmann, "Challenges and solution technologies for the integration of variable renewable energy sources - a review," *Renewable Energy*, vol. 145, pp. 2271–2285, 2020.

44. "IRENA (2018) global energy transformation, a roadmap to 2050," International Renewable Energy Agency, Abu Dhabi, Tech. Rep., 2018.

45. R. Dugan and T. McDermott, "Distributed generation," *IEEE Industry Applications Magazine*, vol. 8, no. 2, pp. 19–25, 2002.

46. W. El-Khattam and M. M. Salama, "Distributed generation technologies, definitions and benefits," *Electric power systems research*, vol. 71, no. 2, pp. 119–128, 2004.

47. D. Xu and A. A. Girgis, "Optimal load shedding strategy in power systems with distributed generation," in *2001 IEEE Power Engineering Society Winter Meeting. Conference Proceedings (Cat. No. 01CH37194)*, vol. 2. IEEE, 2001, pp. 788–793.

48. F. L. Alvarado, "Locational aspects of distributed generation," in *2001 IEEE Power Engineering Society Winter Meeting. Conference Proceedings (Cat. No. 01CH37194)*, vol. 1. IEEE, 2001, pp. 140–vol.

49. L. Coles and R. Beck, "Distributed generation can provide an appropriate customer price response to help fix wholesale price volatility," in *2001 IEEE Power Engineering Society Winter Meeting. Conference Proceedings (Cat. No. 01CH37194)*, vol. 1. IEEE, 2001, pp. 141–143.

50. A. Silvestri, A. Berizzi, and S. Buonanno, "Distributed generation planning using genetic algorithms," in *PowerTech Budapest 99. Abstract Records.(Cat. No. 99EX376)*. IEEE, 1999, p. 257.

51. R. Bansal, "Handbook of distributed generation," *Switzerland Valsan SP, Springer International Publishing, Switzerland*, 2017.

52. M. Sechilariu and F. Locment, *Urban DC microgrid: intelligent control and power flow optimization*. Butterworth-Heinemann, 2016.

53. E. Kabalcı, "Hybrid renewable energy systems and microgrids," 2020.

54. D. W. Gao, *Energy storage for sustainable microgrid*. Academic Press, 2015.

55. G. of India, "National framework for promoting energy storage systems," Ministry of Power, New Delhi, Tech. Rep., Aug. 2023.

56. P. IEA, "Global EV outlook 2023," International Eenergy Agency, Tech. Rep., 2023. [Online]. Available: https://www.iea.org/reports/global-ev-outlook-2023

57. J. Larminie and J. Lowry, *Electric vehicle technology explained*. John Wiley & Sons, 2012.

58. IRENA, "Electric vehicles: technology brief," International Renewable Energy Agency, Abu Dhabi, Tech. Rep., 2017.

59. P. Nieuwenhuis, L. Cipcigan, and H. B. Sonder, "The electric vehicle revolution," in *Future Energy*. Elsevier, 2020, pp. 227–243.

60. P. IEA, "Global EV outlook 2022," International Energy Agency, Tech. Rep., 2022. [Online]. Available: https://www.iea.org/reports/global-ev-outlook-2022

61. C. Gschwendtner, S. R. Sinsel, and A. Stephan, "Vehicle-to-X (V2X) implementation: An overview of predominate trial configurations and technical, social and regulatory challenges," *Renewable and Sustainable Energy Reviews*, vol. 145, p. 110977, 2021.

62. E. Everoze, "V2g global roadtrip: Around the world in 50 projects," *Everoze Partners Limited. UKPN001-S-01-I*, 2018.

63. M. H. Khooban, "An optimal non-integer model predictive virtual inertia control in inverter-based modern AC power grids-based V2G technology," *IEEE Transactions on Energy Conversion*, vol. 36, no. 2, pp. 1336–1346, 2021.

64. M. D. Galus, S. Koch, and G. Andersson, "Provision of load frequency control by PHEVs, controllable loads, and a cogeneration unit," *IEEE Transactions on Industrial Electronics*, vol. 58, no. 10, pp. 4568–4582, 2011.

65. M. Datta and T. Senjyu, "Fuzzy control of distributed PV inverters/energy storage systems/electric vehicles for frequency regulation in a large power system," *IEEE Transactions on Smart Grid*, vol. 4, no. 1, pp. 479–488, 2013.

66. S. Vachirasricirikul and I. Ngamroo, "Robust LFC in a smart grid with wind power penetration by coordinated V2G control and frequency controller," *IEEE Transactions on Smart Grid*, vol. 5, no. 1, pp. 371–380, 2014.

67. J. Yang, Z. Zeng, Y. Tang, J. Yan, H. He, and Y. Wu, "Load frequency control in isolated micro-grids with electrical vehicles based on multivariable generalized predictive theory," *Energies*, vol. 8, pp. 2145–2164, 2015.

68. S. Debbarma and A. Dutta, "Utilizing electric vehicles for lfc in restructured power systems using fractional order controller," *IEEE Transactions on Smart Grid*, vol. 8, no. 6, pp. 2554–2564, 2016.

69. M.-H. Khooban, "Secondary load frequency control of time-delay stand-alone microgrids with electric vehicles," *IEEE Transactions on Industrial Electronics*, vol. 65, no. 9, pp. 7416–7422, 2017.

70. A. Oshnoei, R. Khezri, S. M. Muyeen, S. Oshnoei, and F. Blaabjerg, "Automatic generation control incorporating electric vehicles," *Electric Power Components and Systems*, vol. 47, no. 8, pp. 720–732, 2019.

71. G. B. Gharehpetian and S. M. M. Agah, *Distributed generation systems: design, operation and grid integration.* Butterworth-Heinemann, 2017.

72. S. Padhy, S. Panda, and S. Mahapatra, "A modified GWO technique based cascade PI-PD controller for AGC of power systems in presence of plug in electric vehicles," *Engineering Science and Technology, an International Journal*, vol. 20, no. 2, pp. 427–442, 2017.

73. S. Padhy and S. Panda, "A hybrid stochastic fractal search and pattern search technique based cascade PI-PD controller for automatic generation control of multi-source power systems in presence of plug in electric vehicles," *CAAI Transactions on Intelligence Technology*, vol. 2, no. 1, pp. 12–25, 2017.

74. B. Khokhar and K. S. Parmar, "A novel adaptive intelligent MPC scheme for frequency stabilization of a microgrid considering SoC control of EVs," *Applied Energy*, vol. 309, p. 118423, 2022.

75. Khokhar and Parmar, "Utilizing diverse mix of energy storage for lfc performance enhancement of a microgrid: A novel MPC approach," *Applied Energy*, vol. 333, p. 120639, 2023.

76. S. Falahati, S. A. Taher, and M. Shahidehpour, "Grid secondary frequency control by optimized fuzzy control of electric vehicles," *IEEE transactions on smart grid*, vol. 9, no. 6, pp. 5613–5621, 2017.

77. J. Pahasa and I. Ngamroo, "PHEVs bidirectional charging/discharging and SoC control for microgrid frequency stabilization using multiple MPC," *IEEE Transactions on Smart Grid*, vol. 6, no. 2, pp. 526–533, 2014.

78. R. Rana, M. Singh, and S. Mishra, "Design of modified droop controller for frequency support in microgrid using fleet of electric vehicles," *IEEE Transactions on Power Systems*, vol. 32, no. 5, pp. 3627–3636, 2017.

79. M. Singh, P. Kumar, and I. Kar, "Implementation of vehicle to grid infrastructure using fuzzy logic controller," *IEEE Transactions on Smart Grid*, vol. 3, no. 1, pp. 565–577, 2012.

80. S. of Manufacturers of Electric Vehicles, Online, Oct. 2023. [Online]. Available: https://www.smev.in/statistics

81. A. Aznar, S. Belding, K. Bopp, K. Coney, C. Johnson, and O. Zinaman, "Building blocks of electric vehicle deployment: A guide for developing countries," National Renewable Energy Lab.(NREL), Golden, CO (United States), Tech. Rep., 2021.

82. V. Singh, V. Singh, and S. Vaibhav, "Analysis of electric vehicle trends, development and policies in India," *Case Studies on Transport Policy*, Vol. 9, no. 3, pp. 1180–1197, 2021.

83. S. Goel, R. Sharma, and A. K. Rathore, "A review on barrier and challenges of electric vehicle in India and vehicle to grid optimisation," *Transportation Engineering*, p. Vol. 4, pp. 100057, 2021.

84. H. Bevrani, F. Habibi, P. Babahajyani, M. Watanabe, and Y. Mitani, "Intelligent frequency control in an AC microgrid: Online PSO-based fuzzy tuning approach," *IEEE Transactions on Smart Grid*, vol. 3, no. 4, pp. 1935–1944, 2012.

85. A. Hirsch, Y. Parag, and J. Guerrero, "Microgrids: A review of technologies, key drivers, and outstanding issues," *Renewable and sustainable Energy reviews*, vol. 90, pp. 402–411, 2018.

86. N. Hatziargyriou, Ed., *Microgrids: Architectures and Control*, 1st ed. John Wiley and Sons Ltd, 2014.

87. S. A. Roosa, "Introduction to microgrids," in *Fundamentals of Microgrids*. CRC Press, 2020, pp. 1–16.

88. S. Kaur and S. Prashar, "A novel sine cosine algorithm for the solution of unit commitment problem," *International Journal of Science, Engineering and Technology Research (IJSETR)*, vol. 5, no. 12, pp. 3298–3310, 2016.

89. M. A. Hossain, H. R. Pota, M. J. Hossain, and F. Blaabjerg, "Evolution of microgrids with converter-interfaced generations: Challenges and opportunities," *Electrical Power and Energy Systems*, vol. 109, pp. 160–186, 2019.

90. C. Concordia and L. Kirchmayer, "Tie-line power and frequency control of electric power systems-part II," *Transactions of the American Institute of Electrical Engineers. Part III: Power Apparatus and Systems*, vol. 73, no. 2, pp. 133–146, 1954.

91. M. L. Kothari, B. L. Kaul, and J. Nanda, "Automatic generation control of hydrothermal system," *Journal of The Institution of Engineers India, pt. EL2*, vol. 61, pp. 85–91, 1980.

92. J. Nanda, M. L. Kothari, and P. S. Satsangi, "Automatic generation control of an interconnected hydrothermal system in continuous and discrete modes considering generation rate constraints," *Proceedings of the Institution of Electrical Engineers*, vol. 130, no. 1, pp. 17–27, 1983.

93. T. S. Bhatti, A. A. F. Al-Ademi, and N. K. Bansal, "Load-frequency control of isolated wind-diesel-microhydro hybrid power systems (WDMHPS)," *Energy*, vol. 22, no. 5, pp. 461–470, 1997.

94. T. S. B. A. A. F. Al-Ademi and N. K. Bansal, "Load frequency control of isolated wind diesel hybrid power systems," *Energy Conversion and Management*, vol. 38, no. 9, pp. 829–837, 1997.

95. S. A. Papathanassiou and M. P. Papadopoulos, "Dynamic characteristics of autonomous wind–diesel systems," *Renewable Energy*, vol. 23, pp. 293–311, 2001.

96. P. K. Ray, S. R. Mohanty, and N. Kishor, "Proportional integral controller based small-signal analysis of hybrid distributed generation systems," *Energy Conversion and Management*, vol. 52, pp. 1943–1954, 2011.

97. D. C. Das, A. K. Roy, and N. Sinha, "GA based frequency controller for solar thermal-diesel-wind hybrid energy generation/energy storage system," *Electrical Power and Energy Systems*, vol. 43, pp. 262–279, 2012.

98. R. Mohammadikia and M. Aliasghary, "A fractional order fuzzy PID for load frequency control of four-area interconnected power system using

biogeography-based optimization," *International Transactions on Electrical Energy Systems*, vol. 29, no. 2, pp. 1–17, 2018.

99. L. Wang, *Model Predictive Control System Design and Implementation Using MATLAB*, ser. Advances in Industrial Control. Springer-Verlag London, 2009.

100. A. N. Venkat, I. A. Hiskens, J. B. Rawlings, and S. J. Wright, "Distributed MPC strategies with application to power system automatic generation control," *IEEE Transactions on Control Systems Technology*, vol. 16, no. 6, pp. 1192–1206, 2008.

101. T. H. Mohamed, H. Bevrani, A. A. Hassan, and T. Hiyama, "Decentralized model predictive based load frequency control in an interconnected power system," *Energy Conversion and Management*, vol. 52, pp. 1208–1214, 2011.

102. T. H. Mohamed, J. Morel, H. Bevrani, and T. Hiyama, "Model predictive based load frequency control design concerning wind turbines," *Electrical Power and Energy Systems*, vol. 43, pp. 859–867, 2012.

103. A. M. Ersdal, L. Imsland, and K. Uhlen, "Model predictive load-frequency control," *IEEE Trans. on Power Systems*, vol. 31, no. 1, pp. 777–785, 2016.

104. M. Elsisi, M. Soliman, M. A. S. Aboelela, and W. Mansour, "Model predictive control of plug-in hybrid electric vehicles for frequency regulation in a smart grid," *IET Generation, Transmission and Distribution*, vol. 11, no. 16, pp. 3974–3983, 2017.

105. S. Kayalvizhi and D. M. V. Kumar, "Load frequency control of an isolated micro grid using fuzzy adaptive model predictive control," *IEEE Access*, vol. 5, pp. 16 241–16 251, 2017.

106. H. Bevrani, Y. Mitani, K. Tsujia, and H. Bevrani, "Bilateral based robust load frequency control," *Energy Conversion and Management*, vol. 46, no. 7-8, pp. 1129–1146, 2005.

107. H. Shayeghi, H. A. Shayanfar, and O. P. Malik, "Robust decentralized neural networks based lfc in a deregulated power system," *Electric Power Systems Research*, vol. 77, no. 3-4, pp. 241–251, 2007.

108. S. Patra, S. Sen, and G. Ray, "Design of robust load frequency controller: H∞ loop shaping approach," *Electric Power Components and Systems*, vol. 35, no. 7, pp. 769–783, 2007.

109. A. R. Davidson and S. Ushakumari, "H-infinity loop-shaping controller for load frequency control of an uncertain deregulated power system," in *2016 International Conference on Electrical, Electronics, and Optimization Techniques (ICEEOT)*. Chennai, India: IEEE, Mar. 2016.

110. R. A. Davidson and S. Ushakumari, "H-infinity loop-shaping controller for load frequency control of a deregulated power system," *Procedia Technology*, vol. 25, pp. 775–784, 2016.

111. J. Doyle, "Analysis of feedback systems with structured uncertainties," in *IEE Proceedings D-Control Theory and Applications*, vol. 129, no. 6. IET, 1982, pp. 242–250.

112. H. Bevrani, "Robust load frequency controller in a deregulated environment: a μ-synthesis approach," ser. Proceedings of the 1999 IEEE International Conference on Control Applications (Cat. No.99CH36328). Kohala Coast, HI, USA: IEEE, Aug. 1999.

113. T. Shibata, S. Yoneyama, T. Ohtaka, and S. Iwamoto, "Design of load frequency control based on μ-synthesis," ser. IEEE/PES Transmission and Distribution Conference and Exhibition. Yokohama, Japan: IEEE, Oct. 2002.

114. H. Bevrani, Y. Mitani, and K. Tsuji, "Sequential decentralized design of robust load frequency controllers in multiarea power systems," *IFAC Proceedings Volumes*, vol. 36, no. 20, pp. 109–114, 2003.

115. Bevrani, Mitani, and Tsuji, "Sequential design of decentralized load frequency controllers using μ synthesis and analysis," *Energy Conversion and Management*, vol. 45, no. 6, pp. 865–881, 2004.

116. J. C. Doyle, "Structured uncertainty in control system design," in *1985 24th IEEE Conference on Decision and Control*. IEEE, 1985, pp. 260–265.

117. X. Liu, Y. Zhang, and K. Y. Lee, "Coordinated distributed mpc for load frequency control of power system with wind farms," *IEEE Transactions on Industrial Electronics*, vol. 64, no. 6, pp. 5140–5150, 2016.

118. B. W. Bequette, *Process control: modeling, design, and simulation*. Prentice Hall Professional, 2003.

119. W. Tan, "Tuning of PID load frequency controller for power systems," *Energy Conversion and Management*, vol. 50, pp. 1465–1472, 2009.

120. Tan, "Unified tuning of PID load frequency controller for power systems via IMC," *IEEE Transactions on Power Systems*, vol. 25, no. 1, pp. 341–350, 2010.

121. S. Saxena and Y. V. Hote, "Load frequency control in power systems via internal model control scheme and model-order reduction," *IEEE Transactions on Power Systems*, vol. 28, no. 3, pp. 2749–2757, 2013.

122. Saxena and Hote, "Stabilization of perturbed system via IMC: An application to load frequency control," *Control Engineering Practice*, vol. 64, pp. 61–73, 2017.

123. B. Sonker, D. Kumar, and P. Samuel, "Design of two degree of freedom-internal model control configuration for load frequency control using model approximation," *International Journal of Modelling and Simulation*, pp. 1–12, 2018.

124. I. Kasireddy, A. W. Nasir, and A. K. Singh, "IMC based controller design for automatic generation control of multi area power system via simplified decoupling," *International Journal of Control, Automation and Systems*, vol. 16, pp. 994–1010, 2018.

125. A. J. Veronica and N. S. Kumar, "Internal model based load frequency controller design for hybrid microgrid system," *Energy Procedia*, vol. 117, pp. 1032–1039, 2017.

126. A. Annamraju and S. Nandiraju, "Robust frequency control in an autonomous microgrid: A two-stage adaptive fuzzy approach," *Electric Power Components and Systems*, vol. 46, no. 1, pp. 83–94, 2018.

127. M. A. Johnson and M. H. Moradi, Eds., *PID Control: New Identification and Design Methods*. Springer-Verlag London, 2005.

128. R. Vilanova and A. Visioli, Eds., *PID Control in the Third Millennium: Lessons Learned and New Approaches*, ser. Advances in Industrial Control. Springer, 2012.

129. Y. Lee, S. Park, and M. Lee, "PID controller tuning to obtain desired closed-loop responses for cascade control systems," *IFAC Proceedings Volumes*, vol. 31, no. 11, pp. 613–618, 1998.

130. L. Wang, *PID control system design and automatic tuning using MAT-LAB/Simulink*. John Wiley & Sons, 2020.

131. P. Dash, L. C. Saikia, and N. Sinha, "Automatic generation control of multi area thermal system using bat algorithm optimized PD–PID cascade controller," *Electrical Power and Energy Systems*, vol. 68, pp. 364–372, 2015.

132. Dash, C. Saikia, and Sinha, "Flower pollination algorithm optimized PI-PD cascade controller in automatic generation control of a multi-area power system," *Electrical Power and Energy Systems*, vol. 82, pp. 19–28, 2016.

133. V. Veerasamy, N. I. A. Wahab, R. Ramachandran, M. L. Othman, H. Hizam, A. Xavier, R. Irudayaraj, J. M. Guerrero, and J. S. Kumar, "A hankel matrix based reduced order model for stability analysis of hybrid power system using PSO-GSA optimized cascade PI-PD controller for automatic load frequency control," *IEEE Access*, vol. 8, pp. 71 422–71 446, 2017.

134. D. Guha, P. K. Roy, and S. Banerjee, "Maiden application of SSA-optimised CC-TID controller for load frequency control of power systems," *IET Generation, Transmission & Distribution*, vol. 13, no. 7, pp. 1110–1120, 2019.

135. S. Kumari and G. Shankar, "Maiden application of cascade tilt-integral-derivative controller in load frequency control of deregulated power system," *International Transactions on Electrical Energy Systems*, vol. 30, no. 3, pp. 1–23, 2019.

136. S. Manabe, "The non-integer integral and its application to control systems," *Journal of the Institute of Electrical Engineers of Japan*, vol. 80, no. 860, pp. 589–597, 1960.

137. I. Podlubny, "Fractional-order systems and fractional order PID-controllers," *IEEE Transactions on Automatic Control*, vol. 44, no. 1, pp. 208–214, 1999.

138. Podlubny, *Fractional differential equations: an introduction to fractional derivatives, fractional differential equations, to methods of their solution and some of their applications*. Elsevier, 1998.

139. D. Valério and J. S. Da Costa, *An introduction to fractional control*. IET, 2013, vol. 91.

140. Morsali, K. Zare, and M. T. Hagh, "Applying fractional order PID to design TCSC-based damping controller in coordination with automatic generation control of interconnected multi-source power system," *Engineering Science and Technology, an International Journal*, vol. 20, no. 1, pp. 1–17, 2016.

141. M. I. Alomoush, "Load frequency control and automatic generation control using fractional-order controllers," *Electrical engineering*, vol. 91, pp. 357–368, 2010.

142. S. Farook and P. S. Raju, "Decentralized fractional order PID controller for AGC in a multi area deregulated power system," *International Journal of Advances in Electrical and Electronics Engineering*, vol. 1, pp. 317–332, 2012.

143. S. Debbarma, L. C. Saikia, and N. Sinha, "AGC of a multi-area thermal system under deregulated environment using a non-integer controller," *Electric Power Systems Research*, vol. 95, pp. 175–183, 2013.

144. S. Sondhi and Y. V. Hote, "Fractional order PID controller for load frequency control," *Energy Conversion and Management*, vol. 85, pp. 343–353, 2014.

145. I. Chathoth, S. K. Ramdas, and S. T. Krishnan, "Fractional-order proportional-integral-derivative based automatic generation control in deregulated power systems," *Electric Power Components and Systems*, vol. 43, no. 17, pp. 1931–1945, 2015.

146. I. Pan and S. Das, "Fractional-order load-frequency control of interconnected power systems using chaotic multi-objective optimization," *Applied Soft Computing*, vol. 29, pp. 328–344, 2015.

147. R. Lamba, S. K. Singla, and S. Sondhi, "Design of fractional order PID controller for load frequency control in perturbed two area interconnected system," *Electric Power Components and Systems*, vol. 47, no. 11, pp. 998–1011, 2019.

148. J. Morsali, K. Zare, and M. T. Hagh, "Comparative performance evaluation of fractional order controllers in LFC of two-area diverse-unit power system with considering GDB and GRC effects," *Journal of Electrical Systems and Information Technology*, vol. 5, no. 3, pp. 708–722, 2017.

149. A. Delassi, S. Arif, and L. Mokrani, "Load frequency control problem in interconnected power systems using robust fractional $PI^\lambda D$ controller," *Ain Shams Engineering Journal (Elsevier)*, vol. 9, pp. 77–88, 2018.

150. M. Gheisarnejad and M. H. Khooban, "Design an optimal fuzzy fractional proportional integral derivative controller with derivative filter for load frequency control in power systems," *Transactions of the Institute of Measurement and Control*, vol. 41, no. 9, pp. 2563–2581, 2019.

151. M. Araki and H. Taguchi, "Two-degree-of-freedom PID controllers," *International Journal of Control, Automation, and Systems*, vol. 1, no. 4, pp. 401–411, 2003.

152. H. Taguchi and M. Araki, "Two degree of freedom PID controllers - their functions and optimal tuning," *IFAC Digital Control: Past, Present and Future of PID Control*, pp. 1–6, 2000.

153. J. Sanchez, A. Visioli, and S. Dormido, "A two-degree-of-freedom PI controller based on events," *Journal of Process Control*, vol. 21, pp. 639–651, 2011.

154. R. K. Sahu, S. Panda, and U. K. Rout, "DE optimized parallel 2-DOF PID controller for load frequency control of power system with governor dead-band nonlinearity," *Electrical Power and Energy Systems*, vol. 49, pp. 19–33, 2013.

155. S. Debbarma, L. C. Saikia, and N. Sinha, "Robust two-degree-of-freedom controller for automatic generation control of multi-area system," *Electrical Power and Energy Systems*, vol. 63, pp. 878–886, 2014.

156. Debbarma, Saikia, and Sinha, "Automatic generation control using two degree of freedom fractional order PID controller," *Electrical Power and Energy Systems*, vol. 58, pp. 120–129, 2014.

157. P. Dash, L. C. Saikia, and N. Sinha, "Comparison of performances of several cuckoo search algorithm based 2DOF controllers in AGC of multi-area thermal system," *Electrical Power and Energy Systems*, vol. 55, pp. 429–436, 2014.

158. S. Mishra, R. C. Pruthy, and S. Panda, "Design and analysis of 2DOF-PID controller for frequency regulation of multi-microgrid using hybrid dragonfly and pattern search algorithm," *Journal of Control, Automation and Electrical Systems*, vol. 31, pp. 813–827, 2020.

159. A. Rahman, L. C. Saikia, and N. Sinha, "Load frequency control of a hydrothermal system under deregulated environment using biogeography-based optimised three degree-of-freedom integral-derivative controller," *IET Generation, Transmission & Distribution*, vol. 9, no. 15, pp. 2284–2293, 2015.

160. Rahman, Saikia, and Sinha, "Automatic generation control of an unequal four-area thermal system using biogeographybased optimised 3DOF-PID controller," *IET Generation, Transmission & Distribution*, vol. 10, no. 16, pp. 4118–4129, 2016.

161. D. Guha, P. K. Roy, and S. Banerjee, "Optimal tuning of 3 degree-of-freedom proportional-integral-derivative controller for hybrid distributed power system using dragonfly algorithm," *Computers and Electrical Engineering*, vol. 72, pp. 137–153, 2018.

162. J. Mudi, C. K. Shiva, and V. Mukherjee, "Multi-verse optimization algorithm for LFC of power system with imposed nonlinearities using three-degree-of-freedom PID controller," *Iranian Journal of Science and Technology, Transactions of Electrical Engineering*, vol. 43, pp. 837–856, 2019.

163. I. H. Osman and G. Laporte, "Metaheuristics: A bibliography," 1996.

164. A. Kaveh and T. Bakhshpoori, *Metaheuristics: Outlines, MATLAB Codes and Examples*. Springer Nature Switzerland, 2019, ch. 14, pp. 167–177.

165. S. Arora and P. Anand, "Chaotic grasshopper optimization algorithm for global optimization," *Neural Computing and Applications*, vol. 31, pp. 4385–4405, 2019.

166. T. H. Cormen, C. E. Leiserson, R. L. Rivest, and C. Stein, *Introduction to algorithms*. MIT press, 2009.

167. J.-L. Chabert, É. Barbin, J. Borowczyk, M. Guillemot, and A. Michel-Pajus, *A history of algorithms: from the pebble to the microchip*. Springer, 1999, vol. 23.

168. I. Boussaïd, J. Lepagnot, and P. Siarry, "A survey on optimization metaheuristics," *Information sciences*, vol. 237, pp. 82–117, 2013.

169. N. S. Jaddi and S. Abdullah, "Global search in single-solution-based metaheuristics," *Data Technologies and Applications*, 2020.

170. B. Xi, Z. Liu, M. Raghavachari, C. H. Xia, and L. Zhang, "A smart hill-climbing algorithm for application server configuration," in *Proceedings of the 13th international conference on World Wide Web*, 2004, pp. 287–296.

171. P. J. Van Laarhoven and E. H. Aarts, "Simulated annealing," in *Simulated annealing: Theory and applications*. Springer, 1987, pp. 7–15.

172. F. Glover and M. Laguna, "Tabu search," in *Handbook of combinatorial optimization*. Springer, 1998, pp. 2093–2229.

173. P. Rocca, G. Oliveri, and A. Massa, "Differential evolution as applied to electromagnetics," *IEEE Antennas and Propagation Magazine*, vol. 5, no. 1, pp. 38–49, 2011.

174. J. H. Holland, "Genetic algorithms," *Scientific american*, vol. 267, no. 1, pp. 66–73, 1992.

175. J. R. Koza and R. Poli, "Genetic programming," in *Search methodologies*. Springer, 2005, pp. 127–164.

176. I. Rechenberg, "Evolutionsstrategien," in *Simulationsmethoden in der Medizin und Biologie*. Springer, 1978, pp. 83–114.

177. D. B. Fogel and L. J. Fogel, "An introduction to evolutionary programming," in *European conference on artificial evolution*. Springer, 1995, pp. 21–33.

178. D. Simon, "Biogeography-based optimization," *IEEE Transactions on Evolutionary Computation*, vol. 12, no. 6, pp. 702–713, 2009.

179. D. Dasgupta and Z. Michalewicz, *Evolutionary algorithms in engineering applications*. Springer Science & Business Media, 2013.

180. S. Karimkashi and A. A. Kishk, "Invasive weed optimization and its features in electromagnetics," *IEEE transactions on antennas and propagation*, vol. 58, no. 4, pp. 1269–1278, 2010.

181. M. O'Neill and C. Ryan, "Grammatical evolution," *IEEE Transactions on Evolutionary Computation*, vol. 5, no. 4, pp. 349–358, 2001.

182. S. Mirjalili, A. H. Gandomi, S. Z. Mirjalili, S. Saremi, H. Faris, and S. M. Mirjalili, "Salp swarm algorithm: A bio-inspired optimizer for engineering design problems," *Advances in Engineering Software*, Vol. 114, pp. 163–191, 2017.

183. A. H. Gandomi and A. H. Alavi, "Krill herd: A new bio-inspired optimization algorithm," *Communications in Nonlinear Science and Numerical Simulation*, vol. 17, no. 12, pp. 4831–4845, 2012.

184. M. Dorigo, M. Birattari, and T. Stutzle, "Ant colony optimization," *IEEE computational intelligence magazine*, vol. 1, no. 4, pp. 28–39, 2006.

185. X.-S. Yang, "Firefly algorithm, stochastic test functions and design optimisation," *International journal of bio-inspired computation*, vol. 2, no. 2, pp. 78–84, 2010.

186. R. Oftadeh, M. Mahjoob, and M. Shariatpanahi, "A novel meta-heuristic optimization algorithm inspired by group hunting of animals: Hunting search," *Computers & Mathematics with Applications*, vol. 60, no. 7, pp. 2087–2098, 2010.

187. E. Cuevas, M. Cienfuegos, D. Zaldívar, and M. Perez-Cisneros, "A swarm optimization algorithm inspired in the behavior of the social-spider," *Expert Systems with Applications*, vol. 40, no. 16, pp. 6374–6384, 2013.

188. A. Askarzadeh, "Bird mating optimizer: An optimization algorithm inspired by bird mating strategies," *Communications in Nonlinear Science and Numerical Simulation*, vol. 19, no. 4, pp. 1213–1228, 2014.

189. A.-A. A. Mohamed, Y. S. Mohamed, A. A. M. El-Gaafary, and A. M. Hemeida, "Optimal power flow using moth swarm algorithm," *Electric Power Systems Research*, vol. 142, pp. 190–206, 2017.

190. A. Kaveh and N. Farhoudi, "A new optimization method: Dolphin echolocation," *Advances in Engineering Software*, vol. 59, pp. 53–70, 2013.

191. E. Duman, M. Uysal, and A. F. Alkaya, "Migrating birds optimization: A new metaheuristic approach and its performance on quadratic assignment problem," *Information Sciences: an International Journal*, vol. 217, pp. 65–77, 2012.

192. E. Rashedi, H. Nezamabadi-pour, and S. Saryazdi, "GSA: A gravitational search algorithm," *Information Sciences*, vol. 179, no. 13, pp. 2232–2248, 2009.

193. B. Dogan and T. Olmez, "A new metaheuristic for numerical function optimization: Vortex search algorithm," *Information Sciences*, vol. 293, pp. 125–145, 2015.

194. H. Shah-Hosseini, "Principal components analysis by the galaxy-based search algorithm: A novel metaheuristic for continuous optimisation," *International*

Journal of Computational Science and Engineering, vol. 6, no. 1-2, pp. 132–140, 2011.

195. S. Mirjalili and S. Z. M. Hashim, "BMOA: Binary magnetic optimization algorithm," *International Journal of Machine Learning and Computing*, vol. 2, no. 3, pp. 204–208, 2012.

196. B. Javidy, A. Hatamlou, and S. Mirjalili, "Ions motion algorithm for solving optimization problems," *Applied Soft Computing*, vol. 32, pp. 72–79, 2015.

197. Y. Jun Zheng, "Water wave optimization: A new nature-inspired metaheuristic," *Journal of Computers and Operations and Research*, vol. 55, pp. 1–11, 2015.

198. R. V. Rao, V. J. Savsani, and D. P. Vakharia, "Teaching-learning-based optimization: An optimization method for continuous non-linear large scale problems," *Information Sciences: an International Journal*, vol. 183, no. 1, pp. 1–15, 2012.

199. A. Kaveh and A. Dadras, "A novel meta-heuristic optimization algorithm: Thermal exchange optimization," *Advances in Engineering Software*, vol. 110, pp. 69–84, 2017.

200. V. K. Patel and V. J. Savsani, "Heat transfer search (HTS)," *Information Sciences: an International Journal*, vol. 324, no. C, pp. 217–246, 2015.

201. H. Eskandar, A. Sadollah, A. Bahreininejad, and M. Hamdi, "Water cycle algorithm - a novel metaheuristic optimization method for solving constrained engineering optimization problems," *Computers and Structures*, vol. 110, pp. 151–166, 2012.

202. S. Mirjalili and A. Lewis, "The whale optimization algorithm," *Advances in Engineering Software*, vol. 95, pp. 51–67, 2016.

203. A. Biswas, K. K. Mishra, S. Tiwari, and A. K. Misra, "Physics-inspired optimization algorithms: A survey," *Journal of Optimization*, vol. 2013, pp. 1–16, 2013.

204. R. S. Parpinelli and H. S. Lopes, "New inspirations in swarm intelligence: a survey," *International Journal of Bio-Inspired Computation*, vol. 3, no. 1, pp. 1–16, 2011.

205. B. P. Kumar, G. S. Ilango, M. J. B. Reddy, and N. Chilakapati, "Online fault detection and diagnosis in photovoltaic systems using wavelet packets," *IEEE Journal of Photovoltaics*, vol. 8, no. 1, pp. 257–265, 2017.

206. A. H. Sabry, F. H. Nordin, A. H. Sabry, and M. Z. A. Ab Kadir, "Fault detection and diagnosis of industrial robot based on power consumption modeling," *IEEE Transactions on Industrial Electronics*, vol. 67, no. 9, pp. 7929–7940, 2019.

207. Y. Zhao, T. Li, X. Zhang, and C. Zhang, "Artificial intelligence-based fault detection and diagnosis methods for building energy systems: Advantages, challenges and the future," *Renewable and Sustainable Energy Reviews*, vol. 109, pp. 85–101, 2019.

208. A. S. Bubshait, A. Mortezaei, M. G. Simoes, and T. D. C. Busarello, "Power quality enhancement for a grid connected wind turbine energy system," *IEEE Transactions on Industry Applications*, vol. 53, no. 3, pp. 2495–2505, 2017.

209. Y. Naderi, S. H. Hosseini, S. G. Zadeh, B. Mohammadi-Ivatloo, J. C. Vasquez, and J. M. Guerrero, "An overview of power quality enhancement

techniques applied to distributed generation in electrical distribution networks," *Renewable and Sustainable Energy Reviews*, vol. 93, pp. 201–214, 2018.

210. S. K. Bilgundi, R. Sachin, H. Pradeepa, H. Nagesh, and M. Likith Kumar, "Grid power quality enhancement using an ANFIS optimized PI controller for DG," *Protection and Control of Modern Power Systems*, vol. 7, no. 1, p. 3, 2022.

211. B. Zhao, X. Wang, D. Lin, M. M. Calvin, J. C. Morgan, R. Qin, and C. Wang, "Energy management of multiple microgrids based on a system of systems architecture," *IEEE Transactions on Power Systems*, vol. 33, no. 6, pp. 6410–6421, 2018.

212. H. Gao, J. Liu, L. Wang, and Z. Wei, "Decentralized energy management for networked microgrids in future distribution systems," *IEEE Transactions on Power Systems*, vol. 33, no. 4, pp. 3599–3610, 2017.

213. T. Ma, J. Wu, L. Hao, W.-J. Lee, H. Yan, and D. Li, "The optimal structure planning and energy management strategies of smart multi energy systems," *Energy*, vol. 160, pp. 122–141, 2018.

214. P. R. Lolla, S. K. Rangu, K. R. Dhenuvakonda, and A. R. Singh, "A comprehensive review of soft computing algorithms for optimal generation scheduling," *International Journal of Energy Research*, vol. 45, no. 2, pp. 1170–1189, 2021.

215. J. Yang, N. Zhang, C. Kang, and Q. Xia, "Effect of natural gas flow dynamics in robust generation scheduling under wind uncertainty," *IEEE Transactions on Power Systems*, vol. 33, no. 2, pp. 2087–2097, 2017.

216. M. Nazari-Heris, B. Mohammadi-Ivatloo, K. Zare, and P. Siano, "Optimal generation scheduling of large-scale multi-zone combined heat and power systems," *Energy*, vol. 210, p. 118497, 2020.

217. A. Srivastava and D. K. Das, "A new aggrandized class topper optimization algorithm to solve economic load dispatch problem in a power system," *IEEE Transactions on Cybernetics*, vol. 52, no. 6, pp. 4187–4197, 2020.

218. W.-K. Hao, J.-S. Wang, X.-D. Li, M. Wang, and M. Zhang, "Arithmetic optimization algorithm based on elementary function disturbance for solving economic load dispatch problem in power system," *Applied Intelligence*, vol. 52, no. 10, pp. 11 846–11 872, 2022.

219. Y. Jia, Z. Y. Dong, C. Sun, and K. Meng, "Cooperation-based distributed economic mpc for economic load dispatch and load frequency control of interconnected power systems," *IEEE Transactions on Power Systems*, vol. 34, no. 5, pp. 3964–3966, 2019.

220. A. G. Phadke, P. Wall, L. Ding, and V. Terzija, "Improving the performance of power system protection using wide area monitoring systems," *Journal of Modern Power Systems and Clean Energy*, vol. 4, no. 3, pp. 319–331, 2016.

221. X. Liu, M. Shahidehpour, Z. Li, X. Liu, Y. Cao, and Z. Li, "Power system risk assessment in cyber attacks considering the role of protection systems," *IEEE Transactions on Smart Grid*, vol. 8, no. 2, pp. 572–580, 2016.

222. A. G. Phadke and T. Bi, "Phasor measurement units, wams, and their applications in protection and control of power systems," *Journal of Modern Power Systems and Clean Energy*, vol. 6, no. 4, pp. 619–629, 2018.

223. S. Mirjalili, S. M. Mirjalili, and A. Lewis, "Grey wolf optimizer," *Advances in Engineering Software*, vol. 69, pp. 46–61, 2014.

224. A. E. Ezugwu, O. J. Adeleke, A. A. Akinyelu, and S. Viriri, "A conceptual comparison of several metaheuristic algorithms on continuous optimisation problems," *Neural Computing and Applications*, vol. 32, no. 10, pp. 6207–6251, 2020.

225. R. C. Dorf and R. H. Bishop, *Modern control systems.* Pearson Prentice Hall, 2008.

226. M. Matsubara, G. Fujita, T. Shinji, T. Sekine, A. Akisawa, T. Kashiwagi, and R. Yokoyama, "Supply and demand control of dispersed type power sources in micro grid," in *Proceedings of the 13th International Conference on, Intelligent Systems Application to Power Systems.* IEEE, 2005, pp. 67–72.

227. B. Khokhar, S. Dahiya, and K. S. Parmar, "A novel fractional order proportional integral derivative plus second-order derivative controller for load frequency control," *International Journal of Sustainable Energy*, vol. 40, no. 3, pp. 235–252, 2021.

228. S. K. Pandey, N. Kishor, and S. R. Mohanty, "Frequency regulation in hybrid power system using iterative proportional-integral-derivative H-∞ controller," *Electric Power Components and Systems*, vol. 42, no. 2, pp. 132–148, 2014.

229. W. Zhao, L. Wanga, and Z. Zhang, "Atom search optimization and its application to solve a hydrogeologic parameter estimation problem," *Knowledge-Based Systems*, vol. 163, pp. 283–304, 2019.

230. A. Tepljakov, E. Petlenkov, J. Belikov, and J. Finajev, "Fractional-order controller design and digital implementation using fomcon toolbox for matlab," in *2013 IEEE conference on computer aided control system design (CACSD).* IEEE, 2013, pp. 340–345.

231. D. Valerio and J. S. da Costa, "Ninteger: a non-integer control toolbox for MATLAB," in *Proceedings of the First IFAC Workshop on Fractional Differentiation and Applications, Bordeaux, France, 2004*, 2004, pp. 208–213.

232. S. Mirjalili, "SCA: A sine cosine algorithm for solving optimization problems," *Knowledge-Based Systems*, vol. 96, pp. 120–133, 2016.

233. R. Bansal *et al.*, "Handbook of distributed generation," *Electric Power Technologies, Economics and Environmental Impacts*, vol. 11, p. 6330, 2017.

234. Z. Xu, P. Yang, C. Zheng, Y. Zhang, J. Peng, and Z. Zeng, "Analysis on the organization and development of multi-microgrids," *Renewable and Sustainable energy reviews*, vol. 81, pp. 2204–2216, 2018.

235. M. Gheisarnejad and M. H. Khooban, "Secondary load frequency control for multi-microgrids: HiL real-time simulation," *Soft Computing*, vol. 23, pp. 5785–5798, 2018.

236. A. Askarzadeh, "A novel metaheuristic method for solving constrained engineering optimization problems: Crow search algorithm," *Computers and Structures*, vol. 169, pp. 1–12, 2016.

Index

Printed in the United States
by Baker & Taylor Publisher Services